Constructing Reality

Questions of the fundamental nature of matter continue to inspire and engage our imagination. However, the exciting new concepts of strings, supersymmetry, and exotic matter build on ideas that are well known to physicists but mysterious and puzzling to people outside of these research fields.

Covering key conceptual developments from the last century, this book provides a background to the bold ideas and challenges faced by physicists today. Quantum theory and the Standard Model of particles are explained with minimal mathematics, and advanced topics, such as gauge theory and quantum field theory, are put into context. With concise, lucid explanations, this book is an essential guide to the world of particle physics.

JOHN MARBURGER, III, is University Professor of Physics and Electrical Engineering and Vice President for Research at Stony Brook University. Previous to this, he has been Science Advisor to the President of the United States and Director of the Office of Science and Technology Policy, Director of Brookhaven National Laboratory, and President of Stony Brook University. His accomplishments in science policy and administration have been recognized by numerous awards and honors.

Constructing Reality

Quantum Theory and Particle Physics

JOHN MARBURGER, III

Stony Brook University

CAMBRIDGE
UNIVERSITY PRESS

CAMBRIDGE UNIVERSITY PRESS
Cambridge, New York, Melbourne, Madrid, Cape Town,
Singapore, São Paulo, Delhi, Tokyo, Mexico City

Cambridge University Press
The Edinburgh Building, Cambridge CB2 8RU, UK

Published in the United States of America by Cambridge University Press,
New York

www.cambridge.org
Information on this title: www.cambridge.org/9781107004832

© J. Marburger 2011

First published 2011

Printed in the United Kingdom at the University Press, Cambridge

A catalog record for this publication is available from the British Library

Library of Congress Cataloging in Publication data
Marburger, John H. (John Harmen)
Constructing reality : quantum theory and particle physics / John H. Marburger.
   p.   cm.
Includes bibliographical references and index.
ISBN 978-1-107-00483-2 (hardback)
1. Quantum theory.   2. Particles (Nuclear physics)   I. Title.
QC174.12.M3568   2011
530.12–dc22

                                                                    2011000398

ISBN 978-1-107-00483-2 Hardback

Contents

I The nature of things

I wrote this book for my friends who are not physicists, but who are curious about the physical world and willing to invest some effort to understand it. I especially had in mind those who labor to make the work of physics possible – technical workers in other fields, teachers, science-minded public officials – who read popular accounts but are hungry for a "next step" that might give them a firmer grasp of this puzzling material. Physics gives me great pleasure, more from its beauty than from its usefulness, and I regret that my enjoyment should depend on the effort of so many others who do not share it. Here I have tried to ease my sense of guilt by attempting to disclose in ordinary language what modern physics really is about. Many similar accounts exist.[1] In this one, I attempt to demystify the deep ideas as much as possible in a nonmathematical treatment. Some mathematical ideas are inevitable, and these I try to explain. Physics has entered an exciting phase with talk of new dimensions, exotic matter, and mind-boggling events of cosmic scale. These dramatic ideas rest on a solid conceptual framework, a product of the last century that is now old hat for physicists but remains exotic and impenetrable to most others. This framework, quantum theory and the Standard Model of matter, is an intellectual achievement of the highest order and essential for understanding what comes next. My intention here is to provide a reference and a guide to this known but still regrettably unfamiliar world.

Different physicists have different interests, but I think most would agree that the evolution of our field during the twentieth century stirs deep aesthetic feelings. I will try to explain why this is so, but cannot guarantee readers will have the same reaction. My account is not complete, nor faithful to the complex history of these ideas, but attempts a brisk, coherent sketch of the most important concepts and their links as I understand them. It is unhistorical because it assigns interpretations to past work of which its creators could hardly have been aware. It is personal because it presents my own perspective on the subject, which others may find eccentric. I mean it to be a useful as well as a provocative guide that focuses without much ornament on key ideas. Think of it as a quick review of the conceptual framework of modern physics that requires little prior technical knowledge. It does require patience and mental effort, and I recommend reading it sequentially in short segments. After a first reading, the book may serve as a reference for key concepts. I assume the reader has experienced high school algebra, but has forgotten its details. Notes at the end of each chapter support assertions, add information, and point to further reading. They are written for a wider range of readers who want more detail. Such readers may wonder why I chose to present the material this way. After teaching it for years from a more conventional point of view I realized that the relatively straightforward logic of the physics is easily overwhelmed by numerous fascinating historical or mathematical sidelights. Expert-level accounts let the mathematics carry the argument and omit the side issues entirely. I put them in endnotes. My overriding objective is to disclose the interconnectedness and internal logic of modern physical theory.

To be clear at the outset, my aim is to describe the mainstream view of Nature as expressed in the *Copenhagen interpretation of quantum theory*, and the *Standard Model of matter*. I have tried very hard to avoid saying what these theories are *like*, but rather to say what they *are*. This is notoriously difficult for reasons that will become apparent. I cannot make a difficult subject easy, but at least I can reduce the amount of special technical knowledge, especially mathematical knowledge, required to penetrate to the core of the matter. It is not my aim to probe the inadequacies of the Copenhagen interpretation, but to express it in modern form in the spirit if not in the language of its guiding author Niels Bohr.

The central question is: *If we agree that life is more than a dream, that our consciousness dwells in a universe that includes things other than itself, then what is the nature of those things?*

This vague, possibly meaningless question began, at least in Western culture, with philosophy in ancient Greece, and passed into science in the time of Isaac Newton (1642–1727). For many years thereafter it appeared that the philosophers (some of them) had guessed correctly that all is made of little particles, *atoms*, moving in a void (Leucippus, Democritus, minus fifth century). But in the middle of the nineteenth century the accelerating scrutiny of Nature began to reveal a world disturbingly different from what anyone expected.

I.I NATURE DOES NOT CONFORM TO OUR EXPECTATIONS

Physics has an aesthetic aspect which, like poetry, depends on language and on context. Its context includes both philosophy and the history of discovery. The language of physics, as Galileo first insisted, is mathematics.[2] Many who have peeked at ideas like Einstein's relativity come away shaking their heads, convinced they will never understand them without mastering mathematics. That is a mistake. The difficulty of relativity has nothing to do with mathematics. The same is true of quantum theory. These two pillars of twentieth-century physics are conceptually difficult not because they are mathematical, but because Nature is essentially unhuman.

The linguist Noam Chomsky argued that we have a basic semantic structure hardwired in our brains that renders all human languages deeply similar.[3] Such a structure would have evolutionary survival value only if it resembled the physical environment that challenges our existence. So perhaps we have a reasonable picture of how the world works already embedded in our everyday language. The grammar of cause and effect, of action in the course of time, of place and order, all seem inevitably "natural." Our bodies too, as well as our minds, are equipped to see, hear, and feel "real" things. Immanuel Kant thought Euclid's geometry must represent reality because (to oversimplify his argument) it is physically impossible to perceive, and mentally impossible to conceive, any other kind of geometry.[4]

This anthropocentric view is unfortunately mistaken. The hard-wired structures beneath human language and human perception are not reliable evidence for the deep structure of Nature. Relativity and quantum theory are parts of this structure for which humans do not seem to have any built-in instinct. They fail to conform in deep and important ways to our intuitive preconceptions of how Nature should work.

The presence of "relativistic" and "quantum" ideas in the framework of science is the best refutation of the postmodernist claim that this framework has no independent reality, but is rather a product of social negotiation among disputing scientists.[5] On the contrary, the modern theories emerged painfully from a protracted disputation with Nature herself, and in the end Nature won. Perhaps we have not yet captured her subtlety with our imperfect language, but we are singing to Nature's tune, and not to some completely arbitrary composition of the human mind.

Today we have something called the *Standard Model* which has pieces like a child's toy from which all other ordinary matter can be constructed. Each piece has a name (quarks, leptons, bosons, ...) and properties (charge, spin, flavor, ...) which, together with rules of combination, lead to simple recipes for making *nucleons* (protons, neutrons), *chemical atoms* (clusters of nucleons clothed with electrons), and all else. A chart of the Standard Model (below) suggests the familiar periodic table of elements, of which all chemicals are made.[6] Think of Crick and Watson literally piecing together the structure of DNA with models made of carefully machined parts simulating groups of atoms.[7] We and all about us are made, in a sense to be explained, from the parts of the Standard Model.

This intuitive picture is appealing, and it is also seriously misleading. The pieces that physicists call "particles" are not like anything called by this noun in ordinary language. The broad canvas of "space" and "time" on which the Standard Model is portrayed

u	c	t	γ
d	s	b	G
v_e	v_μ	v_τ	Z^o
e	μ	τ	$W^\pm$

FIGURE 1.1 A conventional table of Standard Model pieces. No shading: leptons. Light shading: quarks. Dark shading: bosons. See Chapter 7 for a different arrangement and explanations of the symbols.

resembles human space and time only in a limited human-scale domain. Quantum theory, the very framework for the modern description of Nature, is strange almost beyond belief. As these deep awkwardnesses became part of physics early in the twentieth century, the field once again acquired a philosophical dimension. The philosophy is not much needed to work problems, but it is important for discovery, and it is essential if we are to make sense, upon reflection, of what it is that we do today when we "do" physics.

1.2 EXPLANATION VERSUS DESCRIPTION

Explanation usually means embedding a phenomenon in a more general framework that we accept as evident.[8] Euclid aimed to reduce geometry to a short list of self-evident axioms and definitions of terms. In the same way, physicists aim to reduce complex phenomena to the action of multiple simple, self-evident, mechanisms. From Newton onward, however, the simple mechanisms ceased to be self-evident. They could be described mathematically (the *how* of the mechanism), but they could not be related to a simple intuitive principle (the *why*). This is in sharp contrast to Aristotle's demand that explanation entail knowledge of the purpose or ultimate cause of a phenomenon, a demand that remains embedded in our culture because it is important in human affairs. We tend to explain human action in terms of motive and objective, but these terms are absent from modern science. In the generation prior to Newton, Johannes Kepler (1571–1630) hypothesized a force on the planets to keep them moving around the Sun, and felt the need to postulate a "soul" that caused it.[9] Newton did not care to explain gravity. He simply described its effect.

This unhuman decoupling of explanation from aim, while necessary, is a psychological impediment for many people. The two are linked by an anthropocentric bias that must be overcome for science to progress.

Mathematics is not the primary obstacle to understanding physics. Students beginning to study the subject are nearly always frustrated because phenomena are not "explained" in a way they expect. They learn to manipulate formulas that give results for various situations – swinging pendulums, falling weights – and this suggests that physics is somehow just mathematical manipulation. Beginners find it difficult to relate the formulas to something tangible. Terms like *force, potential energy, electric field*, are names attached to letters in equations. But what *are* these things, really? Knowing math provides no answer. After more or less experience with the formulas, students acquire intuitions about the behavior of whatever it is that is called by these names, but what the names really signify remains elusive.

It turns out, intriguingly, that the lack of "explanation" makes no difference to how physics is used in applications. No one really understands quantum mechanics intuitively, but hundreds of thousands of scientists and engineers use it in their daily work. The ability to analyze the questions quantum theory was designed to answer does not satisfy our hunger for deeper explanations.

1.3 PHYSICISTS KEEP TRYING TO EXPLAIN THE "UNEXPLAINED"

Its reticence toward explanation has encouraged a rather lifeless view of physics. As the nineteenth century turned, some philosophers embraced positivistic notions about knowledge that discarded concepts that were not rooted in some firm encounter with the commonsensical "real world." The success of "physics without explanations" suggested that attempts to explain were fruitless, and that science should be rid of such baggage. At its worst, this movement doubted the existence of atoms because they could not be seen.[10] At its best, it supported Heisenberg's search for a new atomic mechanics that would depend only on features of atoms that *could* be seen.[11] Some people still speak of scientific formulas as if they were no more than concise summaries of many direct observations, as opposed to statements about the behavior of abstract features of reality, like force and energy, that cannot be visualized.[12] In this view, physics is just a way of arranging experimental results systematically, and the elaborate theoretical structures are only mnemonic devices for the data.

Physicists themselves, however, and especially those who work at the frontier, despite all admonitions from philosophers, seem to believe in the reality of the things their equations describe. They are encouraged in this belief by the great value it has for discovery. In a symposium in 1998 at Stony Brook University, philosopher Bas van Fraasen asked why physicists believe Nature has to obey symmetry laws. I said that "it wins them Nobel prizes!" Throughout the twentieth century, physicists' conviction that abstract entities such as "fields of force" can be "explained" has been influenced by an extraordinary chain of events that I will now endeavor to describe.

NOTES

1. *Similar accounts.* At the same technical level as this book, the excellent account by Crease and Mann (Crease and Mann, 1996) follows the development of modern physics through the contributions of its leading scientists. A more technical presentation by one such scientist, full of insights of interest to the nonspecialist, is Abraham Pais's *Inward Bound* (Pais, 1986). Pais has authored important and well-documented biographies of Einstein, Bohr and others, cited in the References. Helge Kragh's *Quantum Generations* (Kragh, 1999) is a good nontechnical survey that places the subject in a broader social context. On the interpretation of quantum mechanics, David Lindley's *Where Does the Weirdness Go?* (Lindley, 1996) is a lucid account for a general audience. Other references are cited in the notes following each chapter below. *The Whole Shebang: A State-of-the-Universe(s) Report* by Timothy Ferris (Ferris, 1998) gives a snapshot in nontechnical terms of current topics, especially in cosmology, not covered in this book. All these accounts are broader and more general than the present work, which focuses narrowly on the Standard Model and the quantum world view, and not on the sweep of discovery or the state of

knowledge of the entire physical universe. Brian Greene's *The Elegant Universe* (Greene, 1999) is a good popular account of *string theory*, the current most promising attempt to resolve incompatibilities between our current understanding of gravity and the other forces in Nature.

2. *Galileo on the language of physics.* "Philosophy is written in this grand book, the universe, which stands continually open to our gaze. But the book cannot be understood unless one first learns to comprehend the language and read the letters in which it is composed. It is written in the language of mathematics, and its characters are triangles, circles, and other geometric figures without which it is humanly impossible to understand a single word of it; without these, one wanders about in a dark labyrinth." Galileo Galilei, *The Assayer* (1623), translation by Stillman Drake. Reprinted in Drake (1957).

3. *Chomsky on hardwired linguistic structure.* "... the child has an innate theory of potential structural descriptions that is sufficiently rich and fully developed so that he is able to determine, from a real situation in which a signal occurs, which structural descriptions may be appropriate to this signal, and also that he is able to do this in part in advance of any assumption as to the linguistic structure of this signal." *Aspects of the Theory of Syntax* (Chomsky, 1965).

4. *Kant's view of geometry.* "... the space of the geometer is exactly the form of sensuous intuition which we find *a priori* in us, and contains the ground of the possibility of all external appearances (according to their form); and the latter must necessarily and most rigorously agree with the propositions of the geometer, which he draws, not from any fictitious concept, but from the subjective basis of all external appearances which is sensibility itself." *Prolegomena to Any Future Metaphysics* (Kant, 1783).

5. *Social status of scientific reality.* The sharpest statements of the postmodernist claim are made by its critics: " ... science is a highly elaborated set of conventions brought forth by one particular culture (our own) in the circumstances of one particular historical period; thus it is not, as the standard view would have it, a body of knowledge and testable conjecture concerning the 'real' world. It is a *discourse*, devised by and for one specialized 'interpretive community,' under terms created by the complex net of social circumstance, political opinion, economic incentive, and ideological climate that constitutes the ineluctable human environment of the scientist. Thus, orthodox science

is but one discursive community among the many that now exist and that have existed historically. Consequently its truth claims are irreducibly self-referential, in that they can be upheld only by appeal to the standards that define the 'scientific community' and distinguish it from other social formations" (Gross and Levitt, 1994). In striving for clarity, Gross and Levitt have excluded from this statement the essential socio-political aspects of the postmodernist case that make it comprehensible. But that is another story.

6. *Standard Model wall chart.* See the figures in Chapter 7 below. The Contemporary Physics Education Project website (www.cpepweb.org) has a popular version that contains more information.

7. *The DNA model of Crick and Watson.* "The brightly shining metal plates were . . . immediately used to make a model in which for the first time all the DNA components were present. In about an hour I had arranged the atoms in positions which satisfied both the X-ray data and the laws of stereochemistry. The resulting helix was right-handed with the two chains running in opposite directions" (Watson, 1968).

8. *"Explanation."* What is "evident" may not be familiar. "What scientific explanation, especially theoretical explanation, aims at is not this intuitive and highly subjective kind of understanding [reduction to the merely familiar], but an objective kind of insight that is achieved by a systematic unification, by exhibiting the phenomena as manifestations of common underlying structures and processes that conform to specific, testable, basic principles" (Hempel, 1966).

9. *Kepler on the origin of the forces that move the planets.* Kepler had inferred three famous "laws" from careful observations of planetary orbits by his predecessor Tycho Brahe (1546–1601). I. The planets move in ellipses with the Sun at one focus. II. A line from the Sun to a planet sweeps out an area as the planet moves which is proportional to the elapsed time of the movement. III. The square of the orbital period for any planet is proportional to the cube of the size of its orbit. Kepler was hard put to explain how the whole system operated: ". . . I admit a soul in the body of the sun as the overseer of the rotation of the sun and as the superintendent of the movement of the whole world." ". . . the philosophers have commented upon the intelligences, which draw forth the celestial movements out of themselves as out of a commentary, which employ consent, will, love, self-understanding, and lastly command; the soul or motor souls of mine are of a

lower family and bring in only an impetus – as if a certain matter of movement – by a uniform contention of forces, without the work of mind. But they find the laws, or figure, of their movements in their own bodies, which have been conformed to Mind – not their own but the Creator's – in the very beginning of the world and attuned to effecting such movements" (Kepler, 1618).

10. *Doubting the existence of atoms.* "However well fitted atomic theories may be to reproduce certain groups of facts, the physical inquirer who has laid to heart Newton's rules will only admit those theories as *provisional* helps, and will strive to attain, in some more natural way, a satisfactory substitute" (Mach, 1893).

11. *Heisenberg's search for a new atomic mechanics.* The complete abstract of Heisenberg's groundbreaking paper *Quantum-Theoretical Re-interpretation of Kinematic and Mechanical Relations* reads: "The present paper seeks to establish a basis for theoretical quantum mechanics founded exclusively upon relationships between quantities which in principle are observable" (Heisenberg, 1925).

12. *Formulas as summaries of experimental data.* This attitude was frequently expressed by the positivist anti-atomists. See Pullman (1998). Wilhelm Ostwald's declaration of 1895 is an example: "To establish relations between realities, that is to say, tangible and concrete quantities, that is science's responsibility, and science fails to meet it when it espouses a more or less hypothetical image."

2 Matter and motion in space and time

2.1 BERNHARD RIEMANN SPECULATES ON THE EMPIRICAL NATURE OF GEOMETRY

Our story begins with curiosity about the precise shape of the Earth, which is interesting for the direct evidence it could give that Earth rotates in space. Newton pointed out that the pull of gravity (*toward a center*) against the centrifugal force of a spinning Earth (*away from the axis* of rotation) will swell the Earth at its equator, distorting it from a perfect sphere. The effect is small, and at the time no one could look at Earth from space to check its shape.[1]

An ancient estimate of Earth's curvature measured shadows at noon cast by towers separated by a known distance along a line of longitude.[2] This looks too crude to capture the subtle shape Newton predicted. Is there a way of using measurements *completely confined to Earth's surface* to infer topographical distortions? The question caught the fancy of one of the greatest mathematicians of all time, Carl Friedrich Gauss (1777–1855), who created a new subfield of geometry to deal with it.[3] Gauss developed formulas that related measurements on the surface to the three-dimensional swelling of an object. To see how this is possible, imagine a dome-shaped mountain encircled by a road with a second intersecting road that goes straight over the summit, up one side and down the other. The length of the first road is the circumference of a circle – that is, *pi* or π (= 3.14159...) times its diameter – but the length of the second road is *longer* than the circle's diameter (which tunnels through the base of the mountain). So the ratio of the circumference at the base to the diameter measured on the surface over the summit is *less* than π. The deviation of this ratio from π is a measure of the curvature of the dome. (Not the most useful measure. See Note 3.) The flatter the dome, the closer the ratio is to π.

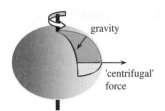

FIGURE 2.1 How can you measure the actual shape of Earth without leaving the surface?

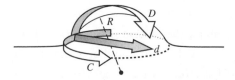

FIGURE 2.2 The circumference C is π times the tunnel route d, and less than π times the over-the-top route D. You can measure C and D without tunneling, and the departure of their ratio from π is related to the radius of curvature R of the dome.[3]

It is remarkable that you can determine the shape of the Earth by making measurements confined entirely to its surface. Gauss's procedure can in principle infer the shape of each bump and dimple in the countryside by strictly local measurements, keeping your nose to the ground, so to speak, without any need of standing back and viewing the whole scene.

Gauss's student, Bernhard Riemann (1826–66), built on these ideas. It occurred to Riemann that our perception of geometry in the three-dimensional world in which we seem to live might be misleading as to its true nature, just as casual observations made on a small part of the surface of the Earth give a mistaken impression of flatness.[4] Since Gauss had shown what to look for in *surface* measurements to tell us if we reside on a curved Earth, there might be similar measurements we can make in the *volume* of our three-dimensional space to tell us if we are perhaps living in a world that has distortions to which our poor human perception is blind. In such a world, some of the geometrical theorems we learned in school would disagree slightly with actual measurements.

In retrospect, there are two important insights here. First, that the axioms of Euclid's geometry cannot be assumed *a priori* to represent reality *à la* Kant, but must be tested by actual measurements in our world. Second, that the world we live in may have additional properties, even additional dimensions, beyond our perception that might make their presence known indirectly. Geometry, in other words, is part of physics – an empirical science rather than a subject, as Plato once supposed, whose truths may be perceived by the mind alone unhindered by gross perception.

Riemann's ideas had no immediate impact on science because no one then knew what aspect of reality might be affected by unseen distortions. Science found little use for the clever mathematical machinery that led Riemann to his insights. That was to change, but not before physicists learned much more about nature. Relativity and quantum mechanics both introduced new geometrical features in our conception of reality.

2.2 THE WORK OF PHYSICS

Science has always progressed in gradually expanding islands of understanding scattered in the sea of ignorance. The island of physics that Newton explored involved massive bodies like planets or objects moving in Earth's gravity.[5] Other disconnected islands in Newton's time concerned light, electric force, and magnetic force. Light is what we see. Electric force makes plastic packing beads stick to our fingers. Magnetic force directs compass needles, turns motors, and empowers magnets to attach notes to the refrigerator door.

Not until the beginning of the nineteenth century did a coherent picture of these diverse phenomena begin to take shape, and by the end of the century a remarkable synthesis had emerged that swept them all into a single framework: *Maxwell's theory of electromagnetism*. This synthesis, accomplished in 1864 by James Clerk Maxwell (1831–79), is a pivotal development in the history of science.[6] Only the discovery of the quantum framework of Nature, itself stimulated by Maxwell's theory, has had greater intellectual impact. Like Newton's description of gravity, Maxwell's description of electricity, magnetism, and light was mathematical and referred to quantities that could no more be visualized than gravitational force.

To appreciate the impact of Maxwell's theory, and the responses to it, you need to know something more about the aim of physical theory. It has two parts:[7]

Motion linked to forces. The first part tells very generally how material objects – think *particles* – move when forced. In the pre-quantum idea of Nature they move simply: if there is no force, particles move in straight lines with constant speed (the *First Law of Motion*); physicists say constant *velocity*, a concept that includes both speed and direction. The effect of a force is to accelerate or to deflect this steady motion. In particular the *acceleration* (rate of change of velocity) is proportional to and in the same direction as the force. For large objects, force can be measured independently – perhaps by the stretch of a coiled spring – so the statement of the Law really does relate two distinct phenomena (force and acceleration). Moreover, the forced acceleration is inversely proportional to a property of the particle that is best called *inertial mass* (this is the *Second Law of Motion*, the famous $F = ma$).

Forces linked to matter. The second part tells how forces are related to the positions and velocities of objects (or of their component particles), and to properties of the particles that I will call *charges*. Thus the electric force exerted by one particle on another is proportional to their *electric charges*, magnetic force is proportional to *magnetic charge* (physicists call it *magnetic pole strength*), and gravitational force is proportional to *gravitational charge*.[8] Because gravitational charge seems to be proportional to inertial mass, physicists call this charge *gravitational mass*. The second part of physics is obviously linked to the first part, since the forces depend on where the particles are, and the particles move in response to the forces. Since the forces are proportional to the charges, it is efficient to frame the theory in terms of *ratios of force to charge*. Thus the *electric field* is the electric force per unit of electric charge. The actual force on a charge at some location is then the field at that location times the magnitude of the charge. Wherever in space you place a charge, it will feel a force (possibly zero), and the word "field" captures the pervasive quality of that which causes the force – the *force field*. It has a direction as well as a magnitude at each point.

As language evolves it carries forward a legacy of ignorance. Words like *mass* and *charge* have come to refer confusingly to several different concepts. Here we distinguish two meanings of "mass," *inertial* and *gravitational*. One has to do with motion in response to *any* force and the other to the property of matter that generates specifically the gravitational force. One of Newton's most important discoveries (prefigured by Galileo at the tower of Pisa)[9] was that the gravitational charge is proportional to the inertial mass that appears in the Second Law of Motion. This proportionality later became an important clue for Einstein when he invented an entirely new theory of gravity that encompassed Newton's formula and changed forever how physicists view their work.

Physicists do two kinds of work: *The work of discovery* searches for all the constituents of Nature and the laws that govern them. Today this enterprise includes *particle physics*, which is the search for fundamental constituents, and other fields such as nuclear physics, condensed matter physics, chemistry, and biology whose scientists search for new things constructed from the fundamental particles. Knowing about the constituent parts and their forces does not suffice to predict all the phenomena that may evolve from them, the possible arrangements into complex objects being far too numerous.[10]

The work of prediction nevertheless employs the laws to attempt to determine how given configurations of matter behave. It is the basis for all applications of physics, and essential for the first kind of work. The laws of motion govern, in principle, the behavior of all phenomena.[11] Nature, however, conceals a marvelous richness beneath the simple rules. New phenomena may be discovered through observations in the laboratory, or alternatively by scrutinizing logical consequences of the rules themselves. Theorists do the latter, experimenters the former, and normally they work together to unravel the deep implications of the deceptively simple laws of Nature.

Newton's laws. In pre-quantum physics the Second Law of Motion gives the acceleration of a particle once the force is known. Since acceleration is the rate of change of velocity, you can find the velocity at any time if you know the velocity just at the beginning and the acceleration throughout the motion. And since velocity is the rate of change of distance traveled, you can find the distance – or position in space – at any time given the velocity and the starting position. So the Second Law contains all the information you need to find the positions of all the particles all the time given only the force law and the starting positions and starting velocities. Isaac Newton invented calculus to solve (among other things) the equations produced by the Second Law.[12] This is the material routinely taught at the very beginning of physics courses in high school or college. It may seem a little cumbersome, but you would not expect to be able to predict the future without some effort. That we can predict anything at all about the future is marvelous.

In the work of prediction the description of Nature separates into two distinct parts, first clearly appreciated by Newton: *initial conditions*, that prescribe from among the infinite variety of possibilities one particular contingent arrangement of Nature's constituent particles and their velocities; and a *law of motion*, each of whose incarnations seems miraculously *to be valid for and independent of any initial arrangement whatever*. The law captures a universal character of motion that makes prediction possible.

Newton's greatest success was to account for the motion of objects – not only planets, moons, and comets, but also the oceanic tides and objects falling near the Earth – mutually influenced by the force of gravity. The other "classical" forces of electricity and magnetism remained somewhat mysterious until the nineteenth century when a burst of experimental and theoretical activity culminated in Maxwell's grand synthesis.

Maxwell's Laws. Maxwell's predecessors – most importantly Michael Faraday – had discovered properties of the electric and magnetic forces, and Maxwell and others developed mathematical expressions to summarize their discoveries.[13] Maxwell added mathematical terms that he found were necessary to make the formulas logically consistent. These terms are ordinarily so small that the experiments of Faraday and others would have overlooked them. Small as they were, the new terms predicted new phenomena (such as electromagnetic radiation) that were subsequently observed. Like Newton's law of gravity, Maxwell's formulas directly summarize simple experiments that anyone can perform on electric and magnetic phenomena. Unlike Newton's laws, their form has never required revision in the light of anything we have discovered since. They possess, moreover, to the eye of the mathematician, a form as beautiful and as natural as a flower.

The physical content of Maxwell's formulas is easy to state, but intricate. They are relations among the rates of change in space and time of electric and magnetic forces in the presence of charges and currents. Hans Christian Oersted, whose observations were developed mathematically by André-Marie Ampère (1820), had shown that an electrical current creates a magnetic field encircling it (1820). Michael Faraday (1831) showed how a *changing* flux of magnetic field creates a concentric circulating electric field in its vicinity. Maxwell (1865) showed that a changing flux of *electric* field must have the same effect on the magnetic field as an electrical current does in order to achieve consistency with charge conservation, which had been demonstrated by Benjamin Franklin (1750). The terms *flux, current,* and *circulating field* have precise definitions that can be related to measurements of distances and forces, so these statements can be translated directly into mathematical relations among quantities that can be measured in the laboratory. Among physicists the resulting *Maxwell's equations* are as famous as $E = mc^2$.[14]

2.3 NEWTON'S UNVISUALIZABLE DESCRIPTION OF NATURE'S ACTION

The idea of reducing Nature to a small set of elements obeying universal rules began in minus fifth century Greece. As far as I can tell, no other ancient culture formed such an objective view of reality. Mathematics had been applied to astronomical phenomena centuries before in Assyrian Babylonia, and a somewhat mystical tradition associating reality with numbers and geometry extended from Pythagoras through Plato into medieval times (and may have had a counterpart in ancient China).[15] But mathematics was not seen as essential for the practical grasp of Nature until the time of Galileo, Descartes, and Johannes Kepler (sixteenth century). Their successor, Isaac Newton (seventeenth century), described the force of gravity with an algebraic formula whose significance could not be grasped in terms of a sensible picture.

Gravity, according to Newton, obeys a law – *the gravitational force between two distant objects is proportional to the product of their masses and to the inverse square of the distance between them* – that can be inferred from observable phenomena (the relation is implicit in Kepler's laws – see Note 9 in chapter 1), but cannot otherwise be explained or visualized. Newton claimed gravity exerts a force across large distances without invoking any cables, rods, or projectiles crossing the intervening space. Later in the eighteenth century the actions of static electricity and of magnetism were found to obey essentially the same law if "masses" are replaced by "charges."[16] Newton's unexplained action at a distance went against the established philosophical grain, and against our instincts as well.[17] "Force" is whatever stretches an ordinary helical spring. How can we visualize such an action without any material link? How can we explain an invisible force like gravity? Newton himself admitted the absurdity of action at a distance, but carried on.

Mechanically inclined people can repair machinery by taking it apart to "see how it works." The transmission of action through the gearing of a steam engine, or the tubing of a hydraulic system, or even the wires and components of an electric circuit, follows a logic of visualizable spatial relationships. Teachers exploit this visualizability to explain technical ideas. But the operation of Newton's forces is not apparent to the eye. To be sure, clever investigators, notably Michael

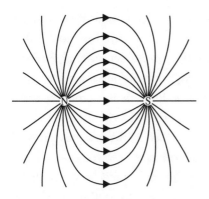

FIGURE 2.3 Faraday's *lines of magnetic force* corresponding to a North–South dipole. A North magnetic pole will experience a force tangent to one of the lines pulling it toward the South pole. The lines can be made visible by placing a horseshoe magnet under a card scattered with iron filings.

Faraday (1791–1867), found ways to depict them. But Faraday's *field lines*, and more intricate machinery invented by Maxwell to picture electromagnetism, are simply not to be seen in Nature.[18] They truly are human inventions to aid the imagination. Iron filings will align themselves with the "lines" of a magnetic field, but the field itself is presumed to exist without the filings. And how does the "force field" exert its influence on the filings? There are no mechanical hooks or strings we can point to that transmit the force from the magnet to the particles of iron. Who has not wondered at the invisible spring between toy magnets that resist our efforts to press them together? Science "explanations" can never remove the sense of mystery surrounding this disembodied force.

2.4 MAXWELL ADDS *FIELDS* TO THE LIST OF THINGS THAT ARE

Maxwell's formulas for electromagnetic forces differed in an essential way from Newton's formula for gravity. Newton's gravitational force is just a postulated cause of acceleration that occurs whenever massive objects (read "gravitationally charged objects") are present. They move toward each other as if they were being pulled by an invisible elastic band. Newton did not know how gravity worked, and thought it was fruitless to speculate. His force formula was indeed just a concise summary of the observed facts of objects moving around on Earth and in the Solar System.

When Maxwell and others examined the formulas that summarized the experiments on electricity and magnetism, they found that *they imply a time lag* between the motion of one electric charge and its influence on a second distant charge. For example, one charged particle might be induced to vibrate, causing the force field it produces to do the same and thus influence the other particle to vibrate too. But the second action follows the first after a delay. Newton's gravitational law, by contrast, implies that a change in distance between two masses translates *immediately* into a change in the force between them, with no time lag. A time lag is puzzling for the following reason: The first particle is losing energy and momentum, the second gaining it. All observations of Nature had shown that energy is conserved – neither gained nor lost in physical processes (the *First Law of Thermodynamics*). If so, where is the energy that is lost by the first particle before the second particle picks it up? Maxwell's answer: *in the electric and magnetic force fields themselves.*

This is a new thing. Maxwell shared Faraday's conception of the electromagnetic fields as real "stuff," capable of carrying properties previously found only in connection with massive particles. The two important properties *energy* and *momentum* are themselves non-intuitive abstract quantities that pop up out of the mathematics of Newton's Second Law. But Newton's Second Law is about the motion of *particles*, the building blocks of Nature. By the end of the nineteenth century most physicists believed in the actual physical existence of atoms.[19] If there is anything "really out there," i.e., outside our minds, then most felt that it should be made of atoms: hard little balls. And here was Maxwell saying that *forces too* can have properties normally associated with tangible particles. In particular, they can carry energy and momentum. The profound effect of Maxwell's work was to add to the list of stuff of which the universe is made. Before Maxwell, it was the atoms of Democritus, moving in void. After Maxwell, it was atoms and *fields of force*, whatever they are. Let us call the fields of Maxwell *dynamic fields* to distinguish them from the lifeless field of gravity that Newton postulated.

We identified two basic parts of physics work, one for the motion of particles (Laws of Motion), the other for the forces among them (Force Laws). Maxwell's work tended to put the second on par with the first. For electromagnetic phenomena, the forces result from dynamic fields that need their own laws of motion. According to this view, Maxwell's formulas are not only a description of how electromagnetic forces arise from charged particles, but also a sort of Law of Motion for electromagnetic fields themselves. Physics work still has two aims: finding laws of motion for every kind of stuff there is, and then solving them to analyze the motion of the stuff. In the reductionist world of Newtonian/Maxwellian physics, everything that *is* is moving stuff.

Physicists use the word *field* to refer to something that exists at every point in space, just as a field of grass ideally has a blade at "every point" of turf. The measure of the "something" – like the height of the grass blade – is said to be a *function* of the location in space.[20] Before Maxwell, you could hope that someday a computer could be made large enough to solve Newton's equations for every atom in existence and thereby predict the future from present knowledge. After Maxwell, the number of equations becomes unmanageably large because the electromagnetic fields have to be computed from initial conditions *at every point in space*. If space is as Euclid imagined, there is an infinity of points even in the smallest volume. If fields are "stuff," then predicting them would seem to require infinite computing power. But perhaps in real space there is not an infinity of points – that is, of places where events can be observed – in each small volume. The small-scale structure of actual space cannot be determined *a priori*. It is a question of physics to be answered by Nature herself, and not by the imagination of mathematicians or philosophers.

2.5 MAXWELL'S IMPACT (1): THE INVARIANT SPEED OF LIGHT

Maxwell's work won immediate and universal acclaim as a profound contribution to the understanding of Nature. But it created conceptual crises that have not yet been completely resolved. Attempts to overcome the problems caused by introducing dynamic fields into the stuff of reality led first to Einstein's theories of relativity and then to quantum theory. Both are needed for our story. Each is in itself an heroic epic, worthy of a Homer or a Milton. My account will be too short, but just long enough, I hope, to make the case for these new theories.

There is first of all the problem of the speed of light. If a time lag occurs between a change in motion of an electric charge (the *cause*) and the response of a second distant charge to the resulting change of electric field (the *effect*), then the ratio of separation to time lag is the speed at which the effect propagates from the cause. Maxwell's equations predicted a value for this speed that turned out to be the same as the then known speed of light (about 186,000 miles or 300 million meters per second) to within experimental accuracy.[21] The implication was that light waves are waves in the electric and magnetic fields. This hypothesis was not confirmed directly until Heinrich Hertz demonstrated electromagnetic waves with the same speed, but much longer wavelengths than light, generated and detected by electric circuits.[22] Oscillating charges in a circuit produce radiation at the oscillation frequency, which is why airlines ask their passengers to turn off electronic devices during take-offs and landings, for fear of interference with the avionics.

Imagine light spreading out ahead of the glowing nose cone of a speeding rocket. We might expect it to move faster than 186,000 miles per second because it adds its own speed to the rocket's. But Maxwell's equations do not say this. They make no reference to the speed of the object emitting the light. They give a speed independent of the motion of the source. Perhaps this is what we should expect if space were filled with a vast ocean of some exotic fluid in which light waves propagate (scientists called it the *ether*). Then if we move *ourselves* through this fluid we expect to see changes in the light speed depending on whether we move along or against the wave direction. But Maxwell's equations make no reference to the speed of the observer either. Our failure to see

FIGURE 2.4 (Left) We might expect light to go faster when emitted by a moving source. (Right) If light is a wave in a medium, it should appear to go faster when an observer is moving toward it in the medium. Neither occurs. Measurements of light speed give the same result regardless of how the source or the observer moves.

an effect in this case could be because the speed of waves in the ether is just so fast that we do not notice the small changes from our own slow motion. Earth itself, however, moves about its axis and the Sun fast enough for its effect on light speed to be easily observable with optical interference techniques. After many careful measurements, no deviation from 186,000 miles per second has ever been detected.[23] No matter how fast either the light emitter or the light observer moves, the speed of light appears to be the same.

It is impossible to visualize any mechanical setup that is consistent with these observed facts. You can try thinking of light as particles shot out from the source, or as waves traveling through a fluid. Neither picture leads to a light speed that remains unchanged when viewed from moving frames. Maxwell's formulas accurately describe a Nature that operates in a nonvisualizable way. This is the first crisis perpetrated upon science by Maxwell's theory.

2.6 EINSTEIN EXPOSES PREJUDICES ABOUT SPACE AND TIME

In the human world speeds are *relative*: standing still on an airplane looks from the ground like moving fast. In Nature, however – very accurately expressed by Maxwell's formulas – at least one speed is apparently not relative in this sense, but *absolute*. The speed of light does *not* change when we measure it from a moving platform. An object viewed as moving at light speed from an airplane looks like moving at the same light speed from the ground.

There is nothing complicated about how we add the velocity of a moving object to the velocity of a moving observer to get a total apparent velocity (velocity entails both speed and direction. To keep things simple imagine the added velocities are parallel). If this process does not work for light or any other substance, there is something very deeply wrong with our thinking. At the turn of the century, this problem boggled the minds of the most brilliant and experienced scientists. To account for the constancy of light speed, they postulated strange effects of the ether on the properties of objects moving through it. But these effects were utterly obscure. Albert Einstein (1879–1955) cut through the confusion with a proposal that was perfectly consistent with the philosophical trend since Newton, but which was received with skepticism and incomprehension by many scientists.

Einstein advocated that we simply accept the fact that the speed of light is constant in any frame of reference moving with its own fixed velocity, and that we *change the rules for adding velocities* to accommodate that fact. He did not try to explain why light speed does not depend on reference frame velocity, he just assumed it is so, *à la* Newton, and moved on.[24]

Speed is just the ratio of distance traveled to time elapsed. We think of time as having absolute significance independent of the speed of any reference platform. You would expect the duration of one second displayed on the watch of an airplane passenger to be the same whether the watch was viewed by the passenger or by an observer in a tower glancing through a window as the plane zooms past. And you would also expect the length of the airplane to be the same whether it is measured by the passenger or from the stationary tower. If these assumptions are true, however, we arrive inevitably at the simple rule for combining velocities that says the observer should measure a different light speed than the passenger, which contradicts Maxwell's empirically based laws.

By accepting the fact that light speed is constant in any frame, Einstein forced a change in how we regard distance and time viewed from relatively moving frames. The mathematics is at the level of high school algebra: Find the simplest rule transforming distances and times from the passenger's frame to the observer's so that light speed is the same in both. Einstein himself provided a simple derivation in a popular account he wrote.[25] The result is that *both sizes (distances) and durations (times) change when transformed from one frame to another.* The stationary observer sees the airplane shrink in the direction of motion, and observes the watch to tick more slowly than his own watch. The formulas that relate durations and distances in one frame of reference to another moving with respect to it are called *Lorentz transformations* because H. A. Lorentz had derived them prior to Einstein from an entirely different, and much more restrictive, point of view.[26]

Time dilation is important in experiments performed with large accelerators because it stretches the brief lifetimes of unstable but rapidly moving particles, like *muons* (Chapter 7), into durations long enough to measure. Neither spatial contraction nor time dilation are noticeable when the objects are moving slowly. They have to be moving near the speed of light, or as physicists say, *relativistically.*

2.7 A DIGRESSION ON $E = mc^2$

What I most want to describe in this chapter is how we came to believe that efforts to explain the simple relations we find in Nature would bear fruit. The thread of this account is knotted with many apparently unrelated discoveries that eventually contributed to a remarkable larger pattern. I vividly recall my first impression of Darwin's *Origin of Species* where descriptions of empirical observations are piled one upon another to create an overwhelming sense of a powerful underlying principle at work. Reading the history of twentieth-century physics creates a similar impression. The side-effects, as it were, of the search for meaning in Nature mount up to reinforce a conviction that we are on the right track. Of all such discoveries, that of Einstein's famous formula was among the deepest, and the least expected.

Newton's Second Law of Motion is about accelerations in response to forces. Acceleration means changing velocity. If the velocity of a particle changes, the relative velocity of the frame in which it is at rest is changing with respect to our "laboratory" frame. So if distances and times have to be transformed whenever relative velocities change, *the law of motion too must be altered* to accommodate Einstein's new ideas. Once again the problem is not mathematically difficult, although it helps to have a little calculus. Suffice it to say that the necessary changes to Newton's formulas can be found easily. Once again, it is not the mathematics that impedes understanding, it is the strangeness of the result. The old familiar Second Law of Motion appears as an approximation to the new one, valid when the object it describes is moving slowly compared with light.

I mentioned earlier that *energy* is not something you can observe directly in the natural environment. It is a quantity derived in physical theory through purely mathematical operations on Newton's Second Law. The formula for energy can be evaluated from measurable properties like masses and velocities. Therefore, if the Second Law is changed to accommodate the constancy of light speed, *then the definition of energy will change too*. That is, the concept of energy is linked to the law of motion. When Einstein examined the relativized Second Law, he found a new expression for the kinetic energy of a particle that did not reduce to zero when the particle was standing still! The algebraic term in the energy formula that remains when all motion ceases is the famous $E = mc^2$. Here m is the inertial mass of the particle and c (for *celerity*) is the speed of light that shows up in Maxwell's equations.

Energy is significant because it is *conserved*, i.e., unchanged during movements described by the law of motion. Einstein's new formula says that the sum of mc^2 and the ordinary kinetic energy (approximately one half the mass times the square of the velocity or $\frac{1}{2}mv^2$) is constant in the absence of forces. This means that if the mass of the object were somehow diminished, then the speed of the object must receive a kick so as to conserve the sum – a BIG kick because the value of c, and therefore of mc^2, is so large. In this way Einstein accounted for the origin of the enormous energies associated with radioactive decay, a complete mystery until then (1905).[27]

Einstein's famous discovery is deeply satisfying because it is so surprising and it is based on such fundamental ideas. It came from the decision to accept as Nature's law the constancy of light speed in Maxwell's theory, and to embed that constancy in Newton's Laws of Motion. Einstein was not working on nuclear theory or radioactivity. Neither he nor Maxwell performed experiments themselves to achieve their results. All they did was try to work out logical kinks the mathematical form of the theory threw into sharp relief. At the outset the mathematical formulas simply summarized the accumulated results of mechanical and electrical experiments. The purely intellectual work required to smooth out the kinks placed the theoretical structure far ahead of experimental observations, many of which were therefore predicted before they were measured. With respect to the new relativistic ideas, experiment has been trying to catch up ever since. For quantum mechanics, the other revolution in twentieth-century science, the roles of theory and experiment were reversed. Experiment rapidly outstripped theory. But that is part of the second set of consequences of Maxwell's revolutionary theory and the subject of the next chapter. Let us return to the story of the revolution in our ideas of space and time.

2.8 MINKOWSKI STRETCHES A NEW CANVAS FOR THE DEPICTION OF NATURE

Newton introduced *forces* acting at a distance without explanation, more or less as shorthand for observations on moving planets. Maxwell showed that the electric and magnetic *force fields* must be as real as massive particles. Einstein said Maxwell's and Newton's formulas together imply that space and time do not act the way we always thought they did. He (and others) found rules for how to go from distances and times in one moving frame of reference to another. Just as the concept of energy had to be changed to accommodate the constancy of light speed, every other concept – momentum, force, electric and magnetic fields, mass densities, etc. – also had to be systematically re-examined and reformulated to deal correctly with relative motion.

That is, the formulas had to be changed that related one quantity to another. The whole body of ideas came to be called *the special theory of relativity*. Our old notions of what is real were transformed into weird concepts that do not match the structures of space, time, and tangibility built into everyday language.

When the mathematician Hermann Minkowski (1864–1909) contemplated these results shortly after Einstein set them forth in 1905, he saw a familiar pattern. The new concept replacing time played a role in the formulas very similar to the new concept replacing distance. Furthermore, the way durations and distances changed in going from one moving frame to another closely resembled the way horizontal and vertical locations change when the reference frame is *rotated*. If a window-washer's platform tilts, the window near the low end appears to be higher to a worker on the platform. The platform has rotated, and the location of the window with respect to it has changed in a definite way. Minkowski noticed that the relativistic transformation formulas were like the formulas describing the window from the tilted perspective of the worker. But in the special theory of relativity, one dimension of the "space" is *time* and the others are the usual dimensions of *length*, *width*, and *height*. This is the origin of the idea that time is a *fourth dimension*. The analogy is very close. It makes itself most evident when the equations are expressed in the language that Bernhard Riemann had advocated for investigating the empirical properties of space.[28]

The four-dimensional language Minkowski advocated for describing events in space-time gives Maxwell's formulas their simplest mathematical form. Calculations are easier when expressed in this language, and results of calculations are easier to interpret. It is as if before Einstein's idea, and Minkowski's language for it, physicists were writing down the formulas in the wrong framework, using the wrong variables and getting clumsy looking mathematical expressions as a result. The new "space" and "time" appear to provide the most natural framework for the description of evolving fields and particles. When relative speeds are slow compared to light speed, these new concepts behave like the old familiar space and time of Newton, and in this regime the new formulas give results that differ negligibly from the old.

The idea of time as a fourth dimension passed quickly into popular culture and language, perhaps too quickly. Today we use the word *dimension* to speak of the number of quantities required to specify something. For example, we may need to ask questions in seven categories to get a good idea of a person's intelligence, in which case we might speak of intelligence as having seven dimensions. We speak of the dimensions of a spreadsheet for accounting, and of the *parameter space* of a machine like a robot arm whose actions need to be programmed. This popular use of the words *dimension* and *space* is somewhat misleading when it comes to the four dimensions of relativistic space-time. Time behaves much more like a true geometrical dimension than the dimensions of experimental psychology or complex machinery. Time and space play such similar roles in the formulas of physics that they mix together when they are transformed to account for the motion of the observer who measures them.[29] That is not to say, however, that time and space are completely interchangeable. The geometrical language favored by Minkowski does single out one among the four dimensions as *time-like*, the others, of course, being *space-like*.

2.9 THE EVOLVING UNIVERSE AS A TAPESTRY OF WORLD-LINES IN SPACE-TIME

Imagine a movie, taken from above, of dancers on a ballroom floor. We shall think of them as the atoms which, according to pre-quantum views, comprise our universe. Now take the film and cut the frames apart and stack them one above another, leaving a little space between each layer. The whole stack becomes a record of the dance in its entirety. If threads were passed through the consecutive positions of each dancer, from one layer to the next, they would make an intricate pattern, like a tapestry. The threads would weave among themselves as the dance progresses from one frame to the next, but the whole dance would be contained within the single fabric.

This is how Minkowski envisioned reality: as a pattern of threads – he called them *world-lines* – traced by each atom within the four-dimensional space-time in which the course of all that exists is manifest. Space is the dance floor, time is the dimension that extends from one frame to the next. The entire volume of space-time encompasses all history as well as all places, so omniscience amounts simply to knowing this pattern of threads. At least it would if all there was to know were the positions of atoms. We are aware since Maxwell's time that we also have to know the values of the electric and magnetic fields at each point and each instant, which seems a vast complication. But for the moment let us focus upon the histories of the atoms.

The first important feature of this picture is that the threads are not broken. They do not begin or end, because atoms in this view are the atoms of Democritus, assumed to be immortal. We should be slightly uncomfortable with this idea, because there is no way we could tell if atoms flickered in and out of existence very briefly between frames, for unobservably short times. But even in the more sophisticated post-quantum view of Nature, what we call *atoms* today are remarkably, if not infallibly, permanent. The atomic threads in the grand tapestry of Existence seem to pass unbroken through countless incarnations in the course of time.

The second important feature of Minkowski's vision is that the dance proceeds according to a remarkably simple set of rules, namely Newton's Second Law of Motion (suitably modified *à la* Einstein to account for finite light speed). How is the Second Law embedded in the fabric of world-threads? Look at Figure 2.5. If there were no force on an atom, its world-line would be straight. Forces cause world-lines to bend. A world-line pointing straight up along the time dimension represents an atom standing still. A world-line straight but tilted represents an atom moving with constant speed in some direction. But an atom accelerating under forces threads a curved world-line, and the more the acceleration the greater (sharper) the curvature.

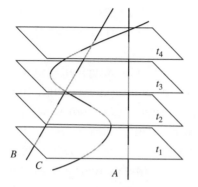

FIGURE 2.5 World-lines in space-time. The particle represented by curve *A* is standing still, *B* is moving with constant velocity, and *C* is accelerating in a wandering pattern. The planes are labeled with the times at which snapshots of the particles are taken.

There is something arbitrary about the distribution of atoms within the space part of space-time. The dancers at some point in their performance may seem chaotically arranged. But with the passage of time a harmony appears. Some dancers avoid each other, others mingle intimately, and a viewer who knows the rules of the dance can always anticipate what will happen next, however the dancers may be arrayed. The rules of motion are reflected in the eternal pattern of world-lines embedded in the space-time picture of the universe.

2.10 EINSTEIN SAYS THE LAWS OF MOTION MUST NOT DEPEND UPON OURSELVES

Einstein insisted that Newton's Second Law be altered to account for the independence of light speed from the arbitrary but steady velocity of a moving reference frame. (Notice that changing frames amounts to tilting the planes in Figure 2.5.) This program of *special relativity* (*special* because the reference frame speed must be constant, not accelerating) swept through all the theories of particle motion and gave them a new and simpler form. Einstein's vision, however, went far deeper than simply reconciling the new features of Maxwell's formulas with the old Newtonian mechanics. He believed the rules governing the dance of atoms should be entirely free of *any* merely contingent features. In particular, he believed the rules should not depend upon where or when or from what perspective they happen to be observed.

(In Figure 2.5 you may choose the planes labeled t_1, t_2, ... to be tilted, warped, or bumpy like rumpled tissues – the world-lines would be the same.)

This is not so easy to arrange. A surveyor will typically express distances and times relative to arbitrary points of reference, like the prime meridian, or Greenwich mean time. These choices clearly depend upon who draws the maps. Einstein wanted the rules of motion to depend only upon *relative* distances, which we might call separations, and *relative* times or durations. Similarly, motions are usually measured with respect to a platform the observer defines to be stationary, such as the (daily rotating!) surface of the planet Earth. So Einstein wanted the rules to depend only on *relative* velocities, as well as relative distances and times. This is why the whole program is called the *theory of relativity*. It is a systematic effort to eliminate the point of view of the observer completely from the laws of motion. The *special* theory of relativity managed to adjust the laws of motion known at the time – Newton's and Maxwell's – to be independent of the position, orientation, historical moment of time, and any *constant* speed, of the observer.

But an observer could be moving with a *nonconstant* speed, jumping up and down for example, or accelerating in some other complicated way like the deck of a tossing ship. And this causes trouble in the theory. Newton's Law is a rule specifically about accelerations. Unlike uniform motion, which we cannot detect, we can always "feel" accelerations (except – and this is a clue – when we are falling). That is the basic experience the Second Law itself summarizes. How can we separate "real" forces from the merely apparent forces inflicted by our own personal, arbitrary, motion?

These apparent forces are familiar. As an automobile (frame of reference) speeds around a curve, the passengers sense a force pulling them to the outside of the curve. They call it *centrifugal force*, but there is nothing there to pull them. They are only feeling the effect of the First Law of Motion, which declares the tendency of bodies (theirs) to move in a straight line. As the auto's path is curving, the passengers' straight trajectories would pass right through its side if they were not pulled along with the auto by friction against the seat and by the seat belts. We attribute the sensation of being pulled outward to a fictitious "centrifugal force." The real force, pulling us in a curve along with the car, is the pressure of the seat belts upon our bodies.

Distinguishing real from apparent forces seems hopeless so long as we express the Rules of the Dance (equations of motion) in terms of the observed acceleration. According to Einstein's program, therefore, *some other formulation of the Second Law of Motion must be found that does not refer to accelerations.* This seems an absurdly radical idea, given that the whole point of Newton's Law is to relate accelerations to forces. Einstein persisted because he was convinced that the deep truths of Nature do not depend in any way upon how we choose to view them. This is the motivation for the *general theory of relativity.*

2.11 A NEW WAY OF THINKING ABOUT THE LAWS OF MOTION

I will not recount the historical steps Einstein followed to his goal. (See chapter 4 of Pais, 1982). But Minkowski's vision of the grand tapestry of world-lines in a four-dimensional space-time, and Riemann's idea that the geometry of our space might be curved without our knowing it, were both essential. What we are looking for, after all, is a way to describe the arching world-lines of the dancing atoms that is independent of an observer's possibly bouncing reference frame. We have seen that acceleration is in this respect a bad criterion for detecting "real" forces. What else can we use to distinguish the "intrinsic" and not merely "apparent" curvature of world-lines? Gauss's work on surface curvature suggests a possibility.

Imagine drawing with a crayon on the surface of a globe. We can scribble any path we like, but we can never draw a perfectly straight line because the underlying globe itself is curved. The straightest thing we can make is part of a *great circle* like a meridian or the equator. This is a path with the *least possible* curvature, namely that of the globe itself. Einstein's grand insight was that perhaps the *observer-independent* curvature of world-lines can be regarded as that of the underlying geometry of the space-time in which the world-lines are inscribed. And that the *apparent* curvature we introduce by our (observer's) jumping about is like the arbitrary scribbling departures from the straightest path we can possibly draw between two points on a globe.

According to this picture, what governs the underlying Rule of the Dance is not the apparent world-line curvature related to acceleration, which depends on the state of the observer, but the intrinsic curvature of the underlying space-time "globe" on which the world-lines wander. That intrinsic curvature must arise from or somehow signal the presence of charges such as gravitational mass or electricity that in Newton's picture produce forces. Any other curvature we happen to observe is just a result of our own (observer's) particular movements, like the artificial centrifugal force. Thus the two-part scheme of physics that relates acceleration to forces and the forces to the charges and disposition of matter is not a good one. Rather we want a scheme that bypasses force and acceleration and directly relates the distribution of charges to the underlying geometry of space-time. *Newton's Second Law would become a statement about world-line geometry*, not about forces and accelerations. *That* rule, if we could find it, would be the ultimate Rule of the Dance.

Einstein failed to find such a rule that applied to all forces. But he did find a more limited rule that applied specifically to the force of gravity, and that in itself was an extraordinary accomplishment.

2.12 AN "EXPLANATION" FOR GRAVITY

Einstein found a way to relate the motion of fields and particles, responding to gravity alone, to the intrinsic geometry of four-dimensional space-time. The idea is that the straightest possible lines in this geometry are the *world-lines actually traced by the particles* and the geometry is determined by the placement of gravitational charges. Thus planets fall freely along the straightest paths in a geometry distorted by the massive influence of the Sun. Who would ever have thought such a description were possible? To judge by his correspondence at the time, even Einstein was amazed. It might have been no more than a mathematical curiosity, except for one thing. *The resulting world-lines differ from what Newton's laws predict*, even when the laws are modified by special relativity. That is, Einstein's *general theory of relativity*, applied to the effects of gravity, predicted observable departures from Newton's Laws. Which theory better fits the facts?

One of the departures had already been observed, and the lack of agreement with Newton's Law was a puzzle for planetary astronomy at the turn of the century. This was the orbit of the planet Mercury which, being close to the Sun and subject to a stronger gravitational force than any other planet, is more sensitive to small deviations from Newton's Laws of Gravity and Motion. In Einstein's theory, the "straightest" planetary path does not exactly close upon itself to make a Keplerian ellipse, but overshoots to trace a rosette, which resembles a slowly rotating ellipse. The rate of rotation predicted by Einstein's theory (43 seconds of arc per century) agreed exactly with the astronomical observations of Mercury after correcting for the influence of the other planets.[30]

Another consequence of Einstein's idea is that not only planets should follow paths whose curvature derives from a space whose geometry is distorted by matter. *Anything* that moves freely through such a space should execute a curved world-line – even a light ray which, being massless, would remain straight according to Newton's Law of Gravity. A light ray curved by the Sun's influence on the geometry it traverses would produce a shift in the apparent location of a star near the Sun, observable during an eclipse. Accurate measurements have confirmed the effect.[31]

In this new theory of gravitation, Einstein took over the mathematical work of Riemann to describe the space-time geometry. Applications of the theory normally proceed in two steps, analogous to the two steps in Newtonian physics. First use the distribution of matter and energy (gravitational charge) to calculate its effect on the geometry of space-time. Then use that information to predict the paths of particles (world-lines) launched with particular initial conditions. This step is purely geometrical, the paths being the analogues of straight lines (called *geodesics*) in the distorted space-time. Since the program of special relativity had already shown that masses are related to energy (through $E = mc^2$), it is necessary to use not only the distribution of mass, but the distribution of *all* forms of energy to determine the space-time geometry. Electric and magnetic fields, mass and kinetic energy, including the energy of motion we associate with heat, all produce distortions in the space-time geometry. If we demand the effects of this distortion to propagate at or less than light speed, then the distortion itself must carry energy by the reasoning I described earlier to attribute energy and momentum to the electromagnetic fields. And this energy too must add further to the distortion, which makes the theory *nonlinear*. Where energy is highly concentrated the two steps must be joined together, the motion of matter-energy being tightly coupled to the dynamical distortion of space-time.

Gravity holds sway over huge distances, and Einstein did not fail to consider that his theory might give information about the shape of the entire cosmos. If the average energy density in our universe is great enough, the geometry of the whole thing could be curved in such a way that the "straightest" paths of particles could eventually close on themselves as they do on the surface of a globe. When applied to the cosmic geometry, the simplest solutions of the new equations were disappointing to Einstein because they implied a dynamic expansion of space itself wherein otherwise fixed objects would appear to be moving with respect to each other. This was long before Edwin Hubble and other astronomers found that indeed the galaxies do appear to be receding (at about $20D$ km/s where D is the distance of the galaxy from Earth in millions of light years). This is but one of numerous tempting byways that threaten to draw us away from the main path of our story.[32]

To summarize: Einstein's grand picture *explains* the force of gravity as an effect of matter and energy upon the geometry of space.

It gives an intuitive flavor to the unexplained cause of acceleration that Newton invoked in his Law of Gravity. And it launched a new odyssey of discovery in which physicists attempted to explain *all* forces by distortions in Nature's geometry.

Skepticism about how much has been "explained" is justified. Energy is a rather abstract idea, and its equivalence to, or effect upon, the curvature of space-time is hardly obvious. What is remarkable is first, that the scheme is so close to what Riemann had imagined and second, that Einstein was not guided primarily by experiment, but by essentially philosophical considerations. He discovered a theory of gravitation more accurate than Newton's by insisting that the laws of motion not depend on arbitrary choices made by the observer. Such fruitfulness of merely philosophical speculation was entirely unexpected, and it changed forever the way physicists approached their subject.

It needs to be said that Einstein's philosophical approach was not uniformly well received by a scientific establishment steeped in positivistic, and therefore experimentally oriented, methods of research. Einstein's Nobel Prize was for work having nothing to do with relativity, and it was a long time coming. In the increasingly anti-Semitic atmosphere of Germany between the World Wars, Einstein's methods were scorned, and Heisenberg himself, who was otherwise acceptable to the authorities, was criticized for teaching them.[33]

2.13 WEYL'S ATTEMPT TO EXPLAIN ELECTROMAGNETISM

As the success of Einstein's scheme emerged, physicists and mathematicians who appreciated it began looking for ways to fit in electromagnetism, at that time the only other known fundamental force in Nature. It was awkward to have one kind of "Rule of the Dance" for the response of particles to gravity, and a completely different one for the response to electromagnetism. If gravity is the result of an intrinsic curvature of space-time, perhaps some other geometrical property of this new four-dimensional world could be identified with electric and

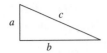

FIGURE 2.6 The sides of a small right triangle in curved space obey $c^2 = g_1 a^2 + g_2 b^2$. In flat space, $g_1 = g_2 = 1$ (Pythagoras's Theorem). The g's define the *metric* of the space.

magnetic forces. Thus began the quest to reduce all forces to geometry. The first idea of how to do this came from the mathematician Hermann Weyl (1885–1955).[34]

Curvature is not the most fundamental property of, nor the best way to describe, the geometry of a multi-dimensional space. It can be derived from a more convenient set of quantities collectively called the *metric* of the space, a concept already employed by Riemann.[35] The numbers defining the metric are coefficients in the analogue of Pythagoras's Theorem that relates the square of a distance traveled to the squares of projections of the distance along the coordinate axes of a reference frame. This sounds rather technical, but the fact that it can be framed in a single sentence suggests that "metric" is not an unduly exotic concept. It relates the distance traveled in some direction to corresponding advances along other directions, just as the hypotenuse of a right triangle can be described as resulting from advances along the two perpendicular legs (Figure 2.6). Curvature is related not to the absolute numbers of the metric, but to how the numbers change as you move about in the space. In Einstein's theory, these numbers play the role of (and for weak gravity are actually proportional to) the Newtonian quantity called *gravitational potential*.

Weyl pointed out that we are free not only to choose the coordinate axes, but also *the measuring scale we use to determine lengths* at each point. Imagine the state of measurement before the nations agreed on a standard meter. There was, so to speak, a French meter, a Swiss meter, and an Italian meter. The numerical stature of an international traveler would grow or shrink at border crossings – a truly artificial effect. But suppose there was an effect that really did change the lengths of things as you moved about (Figure 2.7). Then you would want to distinguish that real intrinsic effect from the artificial one of a changing standard – a changing *gauge*. In any case, you would want the

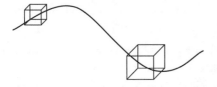

FIGURE 2.7 The size of an object at a point in space could be regarded as a feature of the geometry of the space, analogous to "curvature." In a space that is "flat" with respect to this feature, the size would not change.

law of motion to be *gauge-invariant*: The law should represent reality, no matter what gauge you arbitrarily chose to describe it.

There is no evidence that things change length as they wander from place to place, even in four-dimensional space-time. But in 1918 Weyl worked through the mathematics anyway, and found something interesting. A length scale that changes from place to place can influence the paths of particles just as surely as the changes in curvature that Einstein related to the distribution of energy. In Weyl's theory the scale would vary in the presence of electric charge. He found rules for the force intrinsically associated with the changing gauge (the analogue of the gravity force associated with changing coordinate systems). *And these rules were identical to Maxwell's formulas!*

Weyl, more a mathematician than a physicist (but brilliant in either field), was ecstatic. He sent his discovery off to Einstein, claiming to have *geometrized* electromagnetic theory, and asking him to sponsor its publication. Impressed but skeptical, Einstein sent the work along to be published, but added a note calling attention to its unphysical consequences. The problem was that in space-time a change in length entails a change in duration. Moreover, the changes contemplated by Weyl were not uniquely fixed at each point, but depended upon the path of an object through space. So, remarked Einstein, the oscillation periods of radiating atoms would depend on their histories, and the optical spectra of the elements would blur, contrary to observation.[36]

Weyl's failure, however, was productive. For one thing, mathematical physicists began to realize that the form of Maxwell's equations does closely resemble formulas that arise in the geometry of curved spaces. The idea may not have worked out, but it made the point that Einstein had not exhausted the possibilities. Unfortunately, the metric itself – at least in the four space-time dimensions of Minkowski – did not harbor any additional parameters on which to build a better geometrical explanation of electromagnetism. Furthermore, science was soon to discover the nuclear forces which would require explanations of their own. Things were about to get complicated.

Despite these difficulties, Einstein and co-workers continued to explore other routes to geometrical explanations of all the forces of Nature. None of these attempts at a *unified field theory* bore fruit. Little progress occurred until the revolution of quantum theory had been completed. So now we must turn our attention to this extraordinary development.

2.14 REFLECTIONS ON RIEMANN'S IDEA OF GEOMETRY AS PHYSICS

To Riemann goes the credit for realizing that the geometry of our world may be distorted beneath our notice. Einstein saw that Maxwell's equations implied the need to change our idea of space and time, and launched the program of special relativity. The mathematical form of the result suggested to Minkowski that the proper world was one of four, not three, dimensions. Then Einstein applied Riemann's idea to the geometry of four-dimensional space-time, and identified the distortion of geometry, about which Riemann had speculated, with the intrinsic curvature of world-lines of particles accelerating (falling freely) under the influence of gravity. This concept replaced (for the action of gravity) Newton's idea of relating acceleration to force, and the force to the positions and masses of the gravitating matter. Einstein found explicit formulas that related the local curvature of space-time to the distribution of energy arising from the masses, charges, and motions of the components of matter, including Maxwell's dynamic fields, and worked out the details for various observable phenomena, particularly the orbit of the planet Mercury, and the bending of starlight by the Sun.

The new theory, conceived by untangling the logic of the formulas that described natural phenomena easily accessible to experiment, predicted new phenomena that were subsequently observed. This untangling was not a routine, mechanical process, but a series of creative acts that involved ideas and beliefs about how the universe "really" operates beneath its superficial aspect. The key idea was to ignore the particular accidental arrangement of matter in space, and focus on how that arrangement changes with time. Rules governing change, according to Einstein, should not depend on the arbitrary perspective of an observer. Philosophy creeps into physics through the imperative *should*. Does Nature truly obey this imperative? Subsequent discoveries revealed the need for a much deeper account of the relation between Nature and her observers. Quantum mechanics meets this need, and introduces entirely new features in Nature's intrinsic geometry.

At this point we leave the story of gravity behind. Einstein's geometrical theory changed how we think about force and motion, but its elegant and surprising framework could be adapted to electromagnetism only by introducing exotic new features for which there was no other motivation at the time, and the theory provided no insight at all into quantum phenomena. Physical thought during the ensuing decades produced a new framework – an entirely new language – with which to analyze the microworld. Within that framework a new theoretical structure evolved in parallel with, but apart from, Einstein's theory. Only toward the end of the twentieth century did radical new ideas emerge about how the two might be related, and the unification of gravity with the other forces in Nature is a chief aim of fundamental physics in the new century. My business here is not with these latest speculations, but with that new quantum language – old now to experts – in whose terms physicists believe all future theories must be framed.

NOTES

1. *The shape of the Earth*. Newton's idea is stated as Theorem 16 "That the axes of the planets are less than the diameters drawn perpendicular to the

axes" in Proposition 18 of Book III, *The System of the World* (Newton, 1686). For Earth the relative difference is about 0.3%. A bulge indicates the Earth is spinning about its axis, not fixed with the heavens spinning above. For a modern account of early measurements, see chapter 7 of Wilford (2000).

2. *Ancient estimate of Earth's curvature.* If the distance on the surface between the towers is L, and the difference in latitude θ is measured in *radians* (see Note 3), then Earth's radius is the ratio of L to θ. This method is commonly associated with Eratosthenes (minus second century), preserved in a commentary of Cleomedes (minus first century). See Heath (1932).

3. *Gauss's theory of surfaces.* Appears in his 1827 work translated as *General investigations on curved surfaces* (Gauss, 1827). Gauss's prescription for determining the curvature of a distorted surface leads to the following procedure: Set three pegs to mark a small triangle where you want to know the curvature, and measure its three angles. This requires laying out the sides, which will be curves of the shortest length connecting the pegs and *lying on the surface.* Because of the distortion (e.g., by a bulge) the angles will not add exactly to 180° as would the angles of a flat triangle. Measure the area of the surface enclosed by the triangle. The ratio of that area to the difference of the sum of the angles from 180° equals the square of the radius of an upward bulge within the triangle. (The angles have to be expressed in *radian* measure, in which 180° equals π radians.) This only works if the bulge is spherical. Earth, for example, has two radii at a point on the equator – call them a and b – and Gauss's prescription in this case gives the *geometric mean* of them, namely $\sqrt{ab}$. The deviation from 180° of the sum of angles in a triangle is an example of a violation of Euclid's geometry in a distorted space.

4. *Riemann on geometry.* "... the propositions of geometry are not derivable from general concepts of quantity, but ... those properties by which [our] space is distinguished from other conceivable triply extended magnitudes can be gathered only from experience" (Riemann, 1854). In this famous lecture, Riemann showed how to describe spaces with more than three dimensions, but did not propose that physical space might actually possess more than three. He only suggested that the three we commonly experience might have a geometry different from Euclid's, which is but one of many "conceivable triply extended magnitudes."

5. *The island of physics that Newton understood.* Newton presented his ideas about the nature of the universe in two major published works remarkably

different in their style and content. *The Principia Mathematica* (Newton, 1686) was a Latin treatise modeled after Euclid's *Elements*, setting forth axioms and definitions, and systematically *deducing* from them theorems regarding a host of natural phenomena such as the motions of planets, the tides, the shape of the Earth, etc. The *Opticks* (Newton, 1704), by contrast, appeared first in English and proceeded *inductively* from a series of experiments and questions posed in close connection with a wide range of natural phenomena, not only optics and light. Newton's comprehensive view of nature appears in the famous "*Question 31*. Have not the small Particles of Bodies certain Powers, Virtues, or Forces, by which they act at a distance, not only upon the Rays of Light for reflecting, refracting, and inflecting them, but also upon one another for producing a great Part of the Phaenomena of Nature?"

6. *Maxwell's theory of electromagnetism.* Maxwell expounded his theory in many papers and a famous two-volume book (Maxwell, 1873) which is still in print. The basic relations of electromagnetic theory are explained very simply and clearly in the entirely nonmathematical but stylistically old-fashioned account (Einstein and Infeld, 1938).

7. *The aim of physical theory.* "... we must keep in mind that a physical theory in which we explain phenomena by the motion of small particles consists of two parts; *viz.* 1. the equation of motion of the particles and 2. the rules which represent the forces as determined by the relative positions of the particles, their velocities, electric charges, etc." (Lorentz, 1915). In this book, based on lectures at Columbia University in 1906, H. A. Lorentz summarized his efforts to show how electrical and optical phenomena might be accounted for by the motion of electrons attached to atoms. It is a technical work, but no other book conveys so well the flavor of physics just prior to the revolutions of relativity and quantum theory. Lorentz, for whom Einstein had the greatest admiration, drew attention to the problems of the older theory and the need for new ideas.

8. *Magnetic charge.* Isolated magnetic "north" and "south" polar charges, or *monopoles*, have never been reliably observed in Nature. (A single sighting is reported in Cabrera, 1982.) Magnetic force appears near moving electric charges, as discovered by Oersted in 1820 when he observed a compass needle deflected by a current-carrying wire. Therefore the classical source of magnetism is the *electric current element* (the rate of electric charge flow through a short piece of wire times the length of the piece). The effect of a

small closed current loop is exactly equivalent to that of a *magnetic dipole* with two opposite magnetic charges (north and south) placed close together, the product of their separation and pole strength being equal (up to a coefficient that depends on the measurement units) to the product of the loop area and its current. Magnetic monopoles *could* exist in the framework of quantum theory, and finding one would have important theoretical implications.

9. *Galileo's experiments at Pisa.* Galileo performed experiments demonstrating that when friction is negligible, all bodies accelerate equally, which requires the gravitational force on them to be proportional to their inertial mass. See Cooper (1935), Koestler (1963), and particularly Drake (1981) for the complicated history of these experiments.

10. *The possible combinations of fundamental particles.* Biological molecules (DNA, RNA, proteins, etc.), which are among the most complicated structures ever observed, are built up of combinations of just three chemical elements plus very small admixtures of others. Approximately 62% of the atoms in our bodies are hydrogen, followed by oxygen at more than 26%, carbon at about 10%, and nitrogen less than 2%. The remaining elements are important, but completely dominated by H, O, and C. The huge variety of life processes owes as much to the diversity of *geometrical shapes* of biomolecules as to the diversity of their constituent chemical elements.

11. *The work of prediction.* The apparent determinism of Newton's Laws led the mathematician Laplace to declare that "We may regard the present state of the universe as the effect of its past and the cause of its future. An intellect which at any given moment knew all the forces that animate nature and the mutual positions of the beings that compose it, if this intellect were vast enough to submit the data to analysis, could condense into a single formula the movement of the greatest bodies of the universe and that of the lightest atom: for such an intellect nothing could be uncertain; and the future just like the past would be present before its eyes" (Laplace, 1814). Quoted in Kline (1980, p. 67).

12. *The calculus.* The calculus was invented more or less simultaneously by Newton and by Gottfried Wilhelm Leibniz (1646–1716), and it is Leibniz's notation that we use today. *Integral calculus* finds the area under a given curve A between two vertical limits and an arbitrary horizontal axis. *Differential calculus* recovers the original curve given a second plot B of

the area versus one of the limits. The height of the first curve at a point equals the *rate of change* (slope) of the second curve at the same point. Since Newton's Second Law gives the rate of change of velocity, the integral calculus can be used to find the velocity itself, from which the position may then be found by a second integration.

13. *History of electromagnetism.* A standard history of these developments, with many references to original papers, is Whittaker (1910). Whittaker added a second volume to this work, covering events in the twentieth century, that famously understates Einstein's contributions to the special theory of relativity, but the first volume remains an excellent resource. For an elementary discussion of the laws of electromagnetism, see Einstein and Infeld (1938).

14. *Maxwell's equations.* Sometimes found on engineering students' tee-shirts. They employ mathematical symbols to denote the circulation (*curl*) of an electric ($\mathbf{E}$) or magnetic ($\mathbf{B}$) field, or the diverging flux (*div*) of fields when charges are present. Dots denote rates of change with time, and ρ and $\mathbf{J}$ stand for the densities of electrical charge and current. All quantities are evaluated at a given point in space and time. In a system of units called *Gaussian*, Maxwell's equations for charges moving in a vacuum then become

$$curl\mathbf{E} + \mathbf{B}^{\bullet}/c = 0 \qquad div\mathbf{B} = 0$$
$$curl\mathbf{B} - \mathbf{E}^{\bullet}/c = 4\pi\mathbf{J}/c \qquad div\mathbf{E} = 4\pi\rho$$

If magnetic charges were ever discovered, the zeros in the top row would have to be replaced with terms $4\pi\mathbf{J}_m/c$ and $4\pi\rho_m$, making the equations completely symmetric between electrical and magnetic phenomena (except for the sign difference in the second line). These equations are linked to the motion of a particle with charge q by the *Lorentz force law* $\mathbf{F} = q(\mathbf{E} + \mathbf{v}\times\mathbf{B}/c)$, where $\mathbf{v}\times\mathbf{B}$ is shorthand for a field perpendicular to both the velocity $\mathbf{v}$ and $\mathbf{B}$ with magnitude equal to the area of the parallelogram with sides $\mathbf{v}$ and $\mathbf{B}$. The bold letters denote *vectors* (see Chapter 4).

15. *Mathematics in ancient cultures.* Neugebauer (1957) credits ancient Babylonian astronomy with "... above all, a complete mastery of numerical methods which could immediately be applied to astronomical problems," but no attempt to account for the phenomena, only empirical description. Much the same could be said for the other great astronomical accomplishments of ancient civilizations, including the Chinese and the

Mayan. See volume 3 of Needham (1959), and the long article on Mayan astronomy by Floyd G. Lounsbury in volume 15 of Gillispie (1981).

Aristotle himself summarized the views of the Pythagoreans in Book I of the *Metaphysics*: "... having been brought up in [mathematics] they thought its principles were the principles of all things. Since of these principles numbers are by nature the first, and in numbers they seemed to see many resemblances to the things that exist and come into being – more than in fire and earth and water ... they supposed the elements of numbers to be the elements of all things, and the whole heaven to be a musical scale and a number." A mystical use of numbers also underlies the Chinese *I Ching* or *Book of Changes*, a set of oracular paragraphs keyed to symbols that form all 64 of the six-digit binary numbers. This early use of binary numbers greatly impressed Leibniz, who independently invented them in the West in the eighteenth century. See Cook and Rosemont (1994).

16. *Actions of static electricity and of magnetism.* Charles Augustin Coulomb (1736–1806) carefully measured both electric and magnetic force laws during the 1780s using a torsion balance he invented. His results are embedded in the two Maxwell equations on the right-hand side of the display in Note 14 above. Notice that *all* matter is gravitationally charged, but only *some* matter appears to bear electric or magnetic and other charges. Newton seems to have regarded matter as an inert underlying material whose particles can possess different "Powers, Virtues, or Forces" somewhat analogous to a ball that can be painted with different colors. In the eighteenth century these different "virtues" became quantified by the "charges" of electricity, magnetism, and gravity.

17. *Reception of Newton's "action at a distance."* Philosophers since ancient times had rejected action at a distance, not always for the same reasons. In Newton's time opposition was strongly voiced by Leibniz and Christian Huygens, and even Newton was ambiguous about it (as he mused privately in a letter to Richard Bentley: "It is inconceivable that inanimate brute matter should, without the mediation of something else which is not material, operate upon and affect other matter without mutual contact"). Leibniz attacked Newton's view set forth in the *Principia* in a famous correspondence with Samuel Clarke, an Anglican rector whom Newton had commissioned to translate his "*Optiks*" into Latin. For example, says Leibniz, "A body is never moved naturally, except by another body which touches it and pushes it; after that it continues until it is prevented by another body

which touches it. Any other kind of operation on bodies is either miraculous or imaginary." The exchange is reprinted in Alexander (1956).

18. *Faraday's field lines.* Michael Faraday envisioned the field of force generated by charges as embodied in directed "lines." The lines could be traced, in principle, by noting the direction of the force on tiny test charges due to the other charges present, then smoothly connecting the directions. In this picture the number of lines passing through a small region is proportional to the strength of the force there. See Williams (1965) for details. Maxwell constructed an elaborate system of vortices and "idler wheels" filling the entire vacuum to sustain electromagnetic radiation. His explanation of the system, accompanied by a famous figure of it, is reproduced and discussed in I. J. R. Aitchison's article *The Vacuum and Unification* in Saunders and Brown (1991).

19. *Most physicists believed in atoms.* In the mid-nineteenth century, advances in gas theory by Maxwell, Clausius, and Boltzmann made it possible to relate the actual physical sizes of atoms to *transport properties* such as viscosity and thermal conductivity. The philosophical objections mentioned in the Introduction (Note 10 in Chapter 1) grew toward the end of the century, and then most strongly in continental Europe.

20. *Fields and functions.* Physicists, but not mathematicians, define a *field* as a set of one or more quantities that are defined at each point of a space, or *manifold*, with one or more continuous dimensions. The electric field, for example, is a set of three components of the force exerted on a unit charge (E_x, E_y, E_z) defined at each point of space x, y, z and at each time t. More convenient notation exists (see Chapter 4). A *function* is a relationship between two sets of things. To each *argument* x picked from one set corresponds an object $f(x)$ in the other set. If the things are numbers (which can be represented by lengths along coordinate axes), then the function may be defined by its *graph*, a plot of $f(x)$ versus x. Functions can have multiple arguments, as in the case of the electric field, the x-component of which could be written as $E_x(x, y, z, t)$.

21. *The speed of light.* Already estimated in 1676 by Ole Christensen Römer, who explained certain anomalies in the timing of the appearance of Jupiter's moons as they emerged from behind the planet by supposing that it took time for their light to reach Earth. James Bradley announced a more accurate result in 1729 based on the need to tilt a telescope to view distant

starlight to compensate for the motion of the Earth. The tilt angle (in radians) is very nearly the ratio of the velocity of the Earth to that of light.

22. *Electromagnetic waves.* Hertz performed his experiments in 1886–8 using an electrical circuit with an oscillating current whose frequency he could calculate accurately using the laws of electricity developed by Maxwell and others. He observed a spark in a second, passive, circuit when the first circuit was energized. By moving the second circuit from place to place, he could map out the maxima and minima of the strength of the phenomenon, and thus measure a wavelength. The speed of any wave is the product of its frequency and its wavelength, and Hertz found this to be the known speed of light.

23. *Failure to detect the ether.* The idea of an all-pervasive medium to explain the propagation of light, heat, and even planetary motion, originated with the earliest Greek philosophers. Its modern incarnations took the form of a continuous fluid (Descartes) or densely packed particles (Huygens). Called the *ether*, its existence was more or less taken for granted until careful optical measurements by A. A. Michelson (America's first Nobel Laureate) and E. W. Morley in 1887 failed to detect its effects. The concept has returned in today's quantum theory in several different and interesting guises, but not as the medium in which waves of light are sustained. See Saunders and Brown (1991).

24. *Einstein's principles of special relativity.*

"1. The laws by which the states of physical systems undergo change are not affected, whether these changes of state be referred to the one or the other of two systems of coordinates in uniform translatory motion.

2. Any ray of light moves in the 'stationary' system of co-ordinates with the determined velocity c, whether the ray be emitted by a stationary or by a moving body." (Einstein, 1905a)

25. *Einstein's derivation of transformation laws.* May be found in Einstein (1916), for moving frames. Einstein explains in a preface that "The present book is intended, as far as possible, to give an exact insight into the theory of Relativity to those readers who, from a general scientific and philosophical point of view, are interested in the theory, but who are not conversant with the mathematical apparatus of theoretical physics. The work presumes a standard of education corresponding to that of a

university matriculation examination, and, despite the shortness of the book, a fair amount of patience and force of will on the part of the reader." The derivation in Einstein's appendix 1 requires only simple algebra, but is not easy to follow, and does not convey the exceptional simplicity of the scheme of which the transformation formulas are a part. Unfortunately, that simplicity is only apparent in a (slightly) more advanced mathematical framework. See Mermin (2005) for a lucid introduction to relativity theory.

26. *Spatial contraction and time dilation.* To a stationary observer an object of length L moving along its length with velocity v has a shorter length $L' = L\sqrt{(1-v^2/c^2)}$, a phenomenon called *Lorentz contraction*. A moving clock is observed to run slower by the inverse factor, so the time between ticks would be longer: $T' = T \div \sqrt{(1-v^2/c^2)}$. In this case time is said to be *dilated* in the moving frame. Both phenomena are real, not just apparent. See Mermin (2005).

27. *Energies associated with radioactive decay.* "The mass of a body is a measure of its energy-content; if the energy changes by ε, the mass changes in the same sense by $\varepsilon \div 9{\times}10^{20}$, the energy being measured in ergs, and the mass in grammes. It is not impossible that with bodies whose energy-content is variable to a high degree (*e.g.* with radium salts) the theory may be successfully put to the test" (Einstein, 1905b). Einstein used L in place of ε. The test could not be performed without considerably more knowledge of the structure of nuclei. For details, see Pais (1982, paragraph 7(b)).

28. *Minkowski stretches a new canvas.* Minkowski's 1908 lecture to a nonspecialist audience (Minkowski, 1908) begins with this dramatic paragraph: "The views of space and time which I wish to lay before you have sprung from the soil of experimental physics, and therein lies their strength. They are radical. Henceforth space by itself, and time by itself, are doomed to fade away into mere shadows, and only a kind of union of the two will preserve an independent reality."

29. *"Mixing" of space and time.* Undermines the concept of "simultaneity." Two lightning strikes at points A and B on a railway embankment may seem simultaneous to a stationary observer on the ground, but owing to the finite speed of light the strike nearer the front of a moving train will be seen by a moving observer to come first. "Events which are simultaneous with reference to the embankment are not simultaneous with respect to the train, and vice versa (relativity of simultaneity). Every reference body (co-ordinate system) has its own particular time; unless we are told the

reference body to which the statement of time refers, there is no meaning in a statement of the time of an event" (Einstein, 1916).

30. *The rate of rotation predicted by Einstein's theory.* According to Pais, who knew Einstein well, "This discovery was, I believe, by far the strongest emotional experience in Einstein's scientific life, perhaps in all his life. Nature had spoken to him. He had to be right. 'For a few days,' [he wrote to Paul Ehrenfest] 'I was beside myself with joyous excitement'" (Pais, 1982, p. 253).

31. *Measurements of gravitational stellar shifts.* The most famous such measurements, not at all accurate, were orchestrated by Arthur Eddington during the unusually long total solar eclipse of May 1919 when starlight from the Hyades cluster passed sufficiently near the Sun. Reports that these data supported general relativity made Einstein an international celebrity. The most accurate and reliable measurements observe radio signals from astronomical objects, which are "visible" even in the absence of an eclipse. See Pais (1982).

32. *Receding galaxies.* Recent astronomical discoveries have made cosmology one of the most exciting areas of modern science. Not only are the galaxies receding from one another, but the speed of recession has apparently been *accelerating.* See, for example, Weinberg (1993), Ferris (1998), and Livio (2000).

33. *Einstein's philosophical approach.* Pais's biography of Einstein includes a balanced account of the circumstances of Einstein's Nobel Prize (for the photoelectric effect, not for relativity), and the lukewarm reception of his theories in some circles (Pais, 1982). See also Kragh (1999) for the state of physics in Weimar Germany.

34. *Quest to reduce all forces to geometry.* At the time, the only known forces were gravity and electromagnetism, and Weyl's 1918 approach was the first to unify them in a single geometrical framework. A few years later, Theodor Kaluza proposed an apparently very different unification which introduced a new spatial dimension – a *fifth dimension* (1921). This approach fell out of the mainstream of physics for the next half-century, but is curiously related to the successful post-quantum modification of Weyl's theory. A mathematical step called *dimensional reduction* connects coordinate transformations in five dimensions with gauge transformations in four. Today Kaluza's approach is important in efforts to unify the phenomena of the Standard Model with gravity. See Greene

(1999) for a popular account of these developments. A good but very technical account is O'Raifeartaigh (1997).

35. *Curvature not the most fundamental property.* Nor is the metric the most primitive foundation for geometry. Weyl pioneered a new direction in geometry based on the concept of *parallel transport.* How geometrical properties change from place to place can be described in terms of parameters that determine how directions are altered as they are transported from one point to another along specified paths. These parameters, which themselves can change from point to point, are collectively called *connections* in the space. In Weyl's theory, electromagnetic fields (or rather their potentials) are associated with the connections. Concepts of metric, curvature, and connection are all part of the field of *differential geometry* that began with Gauss (1827) and has since grown steadily in importance to physicists as the intrinsically geometric character of Nature has become apparent.

36. *Weyl's geometrization of electromagnetism.* Thus Einstein: "If this [length of an object dependent on its history] were really so in Nature, chemical elements with spectral-lines of definite frequency could not exist and the relative frequency of two neighboring atoms of the same kind would be different in general. As this is not the case it seems to me that one cannot accept the basic hypothesis of this theory, whose depth and boldness every reader must nevertheless admire." Weyl's original paper is translated in O'Raifeartaigh (1997), where Einstein's criticism and Weyl's response are also reproduced.

3 Reality large and small

3.1 DIGRESSION ON THE QUALITY OF KNOWLEDGE IN A UNIVERSE OF ATOMS

Notice how cerebral all these advances were of Einstein's. With little data and few experiments to guide him, he plumbed the deep logic of the formulas that summarized so much of Nature then known to science. At about the same time, however, triggered by technologies whose power had mounted throughout the nineteenth century, a Pandora's box of new physical phenomena sprang open for which the old formulas failed to account. Subsequent progress in theory required a steady stream of reports from the laboratory to weed out conceptual dead ends and suggest new directions. The new experiments probed matter's fine structure and its behavior at extremes of temperature and pressure. New theories to explain them conceived matter as a concretion of particles held together by electromagnetic forces. The component particles were supposed to be more or less like those of Democritus, but bound electrically to produce composite *atoms* corresponding to each of the chemical elements within Mendeleev's famous *periodic table*.[1]

School children today learn that the elemental atoms consist of lightweight negatively charged *electrons* bound electrically to tiny, but very heavy, positively charged *nuclei*. More than 99.9% of the atom's mass resides in the nucleus, which, however, is a surprising 100,000 times smaller than the diameter of the atom itself. So the electrons must be orbiting the nucleus much as planets circulate far from the Sun. From the viewpoint of the Newton–Maxwell theory, this picture harbors staggering paradoxes, not least how the electrons remain so distant from the tiny nucleus, given the huge inwardly directed electrical force. Unlike planets, orbiting charges would lose energy rapidly through radiation as in Hertz's experiment with electric circuits (see Note 22 in chapter 2) and spiral toward the attracting center. Today, at the scale of the chemical atoms, all paradoxes have been resolved through the view of Nature painstakingly constructed during the first quarter of the twentieth century. It is not an obvious view, and I need to make a few preliminary remarks for the sake of orientation.

I ask that you take seriously for a moment the idea that *everything* is made of tiny bits (not to be confused with the information "bits" that feed computers). I am avoiding the word "atoms" here to emphasize this is merely a thought experiment. This is an extreme but easily visualizable notion imaginatively espoused by the Roman poet Lucretius (minus first century) in his long work *On the Nature of Things*, which elaborated the vision of Leucippus and Democritus. Imagine even that consciousness itself somehow springs from the enormous accumulation of bits that we are. How, in this picture, would we become aware of the world around us? Some little bits must travel as mediators between our bits and the bits in the rest of the world, bringing information about what happens. To carry information, a mediator bit must have been produced or deflected by the world outside, and then it must affect the bits that make up our sensors (eyes, ears ...). That means, however, that the outside world must retain some quivering memory of the influence it exerted to signal its existence. (*To every action there is an equal and opposite reaction* – the third of Newton's three Laws of Motion.) So our seeing requires some faint disturbance in the world we see.

I am not saying the world is truly made of little bits, or what those bits are – this is just a game of the imagination like Lucretius's poem. In this game, sensing an object requires mediating bits to travel from the object to a sensor. Make a source of mediating bits – a "flashlight." Shine it on the object to be viewed. A stream of bits strikes the object, rebounds to leave the thing quivering, and glances into our sensor (eye) where it stirs our awareness. If the object is big enough it will scarcely feel the beam, and our awareness will accurately represent its place and state of motion. If the object is no bigger than the mediating bits, then its course will be deflected, and we must make a mental note to adjust our awareness to account for the disturbance of the act of sensing.

If we know which way the flashlight beam is pointing, and notice the direction from which the mediating bit enters our eye, then we might be able to infer (within some technical limits) where the object was when it deflected the bit. We have "measured" the position of the object. Unfortunately, the encounter with the flashlight bit has sent the object off in a new direction with a new speed, and a subsequent position measurement cannot be used without further information to estimate the original velocity (by dividing the distance between the two positions by the time between measurements). Measuring speed and position simultaneously could therefore be tricky. The disturbance linked to the getting of information could possibly limit what we can glean from a single measurement.

This train of thought makes it not unreasonable that what we demand of physical theory may depend upon how ponderous the observed object is relative to the detecting bits. For large objects we can ignore the whole detection process and rely entirely on our senses without any mental reservations. For really small objects a proper theory had better account for the disturbances caused by the act of measurement. Call the usual theory *macroscopic*, and the new one *microscopic*. The profound role of the observer and the act of observing in microscopic theory contrasts strongly with Einstein's urge to rid theory entirely of arbitrary choices by observers. This does not entirely account for Einstein's later criticism of quantum theory, but he did go off on his own looking for a picture closer to the old macroscopic models.

A noteworthy feature of this hypothetical world of bits is that the only definite empirical knowledge we can have of the universe is somewhat discontinuous. We would experience the external world through isolated brief encounters with messenger bits from outside. It is as if all we knew of the tapestry of world-lines were the knots where the messenger bits intersect with our own. If Nature works this way the apparent continuity of existence is an illusion that may or may not reflect microscopic reality. On the large scale of human action, however, the myriad individual events of sensation blur together like the frames of a moving picture. Our mind fills in the world-lines between the knots.

Not until the end of the nineteenth century did scientists seriously encounter the world of the small. No one had thought much about the limitations on observation imposed by the finite impact of the particles required for sensation. For one thing, you could always imagine a god who could see with a delicacy beneath the reach of mortals, and you could strive to discover the laws as the god would perceive them. For another, no one had yet needed to invoke any physical system small enough to be disturbed by something with a touch as light as, well, light itself. Then in 1897 J.J. Thomson (1856–1940) discovered the *electron*, nearly 2000 times less massive than the least of the 90-odd different kinds of atoms. It was not a chemical atom itself, but, as soon became apparent, it was part of one. Electrons, moreover, turned out to be supremely important in the human world, figuring prominently in all the new technologies that in the ensuing century transformed our way of life. Scientists pieced together the quantum puzzles mostly through analyzing how electrons moved within atoms, and how observers might convince themselves the analysis was correct.

Now I will begin to describe the world view of quantum theory. There is something exceedingly strange about this theory, and it is precisely the origin of that strangeness which is important for the tale I wish to tell. But what is most strange is *not* that measurement disturbs the object measured. Or that a separate theory is needed for small things. Or even that, as it turns out, the best we can do is make statistical statements about Nature, because we can expect the disturbance of measurement to be somewhat random and uncontrollable. I wanted to lay these ideas out now so you can separate them from the things that *really* are strange and unexpected about the quantum world.

3.2 MAXWELL'S IMPACT (2): THE MISMATCH BETWEEN PARTICLES AND FIELDS

Democritus would have been alarmed at the notion that fields of force could be just as real as atoms. *Fields* are defined "everywhere." That is, they can be different at each of the (presumed) infinity of points within any small volume. Near each of these points the electric or magnetic forces can be as strong or as weak as you like. According to Maxwell there will be a correspondingly great or small energy associated with each tiny region.[2] It appears that an arbitrarily small volume of space can hold a huge amount of field energy, and that energy can be distributed within the volume in a huge number of different ways. A single particle, idealized as a "mass point," is said to have three *degrees of freedom*, one for each of the three different directions it can move (up–down, left–right, back–forth). The electromagnetic field is present at each of the infinite number of points in any finite volume *and therefore has an infinite number of degrees of freedom.*

Heat radiation. This has consequences for the thermal properties of light. When you peek into a pottery kiln you see a reddish glow. The hotter the interior walls, the whiter or bluer the color. If light were made of bits you can imagine how everything heats up: The fire beneath the kiln heats or jiggles the atoms of the walls, and they jiggle the bits of light. Maybe the atoms' jiggling converts some of their energy into new bits of light. Pretty soon all the bits come into *thermal equilibrium* with the walls, which means they reach the same temperature as the walls, which means that on the average each bit of light has the same jiggling energy as a typical atom in the wall. In this state, the temperature rise is definitely related to the total heat input to the kiln and the number of bits within, assuming no heat is lost from this perfectly insulated oven. Putting in heat raises the temperature (the ratio of heat input to temperature rise when both are small is called the *specific heat*). If there are lots of light bits to heat up, then the final temperature will be lower than if there were just a few, because the heat is shared out among all the bits when things settle down into thermal equilibrium.

Maxwell says light is not made of bits, but of waves in the continuous electromagnetic field of force. So run through the kiln scenario again with fields instead of bits. When the walls are heated, their atoms jiggle, and their electrons too. Their jiggling shakes off some light waves into the space within the kiln. Or you might think of the fields as latent within this space, awaiting the motion of the brick atoms or the charged electrons and nuclei to which they are attached (as described by Maxwell's formulas). The jiggling charges energize the field as you might flick the end of a jump rope to get it going. In this way the heat input goes into the fields as well as the bricks. But each little part of space within the kiln has an infinity of points at which the fields can be excited. These will suck up all the heat energy, so as things settle down toward thermal equilibrium, each of the huge number of points in space will share a very small amount of the input heat, and the bricks will have given over all but a negligible amount of their energy to the uncountable multitude of jiggling field points. The bricks will never even get warm to the touch, and the light will not look hot. This obviously is not what happens.

So there's the problem. You should not be able to heat up light because its specific heat is too great – it ought to be infinite. But nevertheless we do see "hot light" through the kiln's peek-hole. Something must be wrong.

3.3 PLANCK POSTULATES A RELATION BETWEEN ENERGY AND FREQUENCY

The redness of light escaping from a hot oven can be measured. Use a prism to make a rainbow of the light and plot the intensity at each color with a light meter. Color can be quantified by the vibration frequency of the light waves – lower frequency for red, higher for blue. The result is a graph of light intensity versus wave frequency shown in Figure 3.1. This is a graph of the spectrum of *black body radiation* at that temperature – "black body" because the situation is simplest when the hot walls or atmosphere of the oven have no color of their own to get in the way. If the walls were cold, they would look black. Experimental

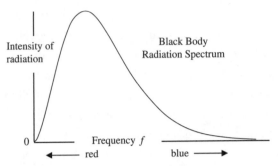

FIGURE 3.1 Planck's black body radiation spectrum has the same shape for every temperature but its peak shifts to higher frequencies (toward blue) as the temperature increases. At the melting point of iron (about 1540°C) most of the spectrum is in the infrared.

physicists measured the black body spectrum very carefully at the end of the nineteenth century.[3] What they saw was low intensity at low frequencies (infrared light), and rising intensity to a maximum depending on the temperature of the oven (more blue for higher temperatures), and then rapid dimming to zero at higher frequencies (toward the ultraviolet). It is basically a humped curve, and it is what you might expect to see if light were bits, not waves. What is striking is the fall-off at high frequencies. It is as if that part of the spectrum were difficult to excite. Apparently *it takes more energy to excite higher frequencies of radiation* than it does to excite lower frequencies. This is a fundamental fact of Nature, but it is not embedded in the summarizing formulas of Newton/Maxwell.

In 1900 Max Planck (1858–1947) made these ideas quantitative (Planck, 1900). Using statistical methods to describe the state of thermodynamic equilibrium, he showed that it was possible to account for the black body spectrum if only one assumed that heat is exchanged between the radiation and the atoms of the walls in *quanta* of energy proportional to the frequency: $E = hf$. The higher the frequency f (measured in *hertz*, or cycles per second) the more energy has to be transferred to excite it. Planck found the value of the proportionality constant h, today called *Planck's constant*, that fit the experimental data. It is a very small number in units that are useful at human scales: $h = 6.6 \times 10^{-34}$ joule seconds. (A joule is the energy required to power a one-watt light bulb for one second.)[4]

This was unexpected. There is nothing in the Newtonian definition of energy for an oscillating system that links it in this way to the vibration frequency.[5] Planck had no model, no theory, and no evidence for his proposition, other than the particular case of black body radiation, where it worked. He produced a formula whose curves fell exactly on the experimental data for all the measured spectra. For a while, no one knew what to make of it.

There the matter stood for several years until Einstein pointed out that Planck's relation might account for several other puzzles in experimental data. Based on a (slightly) improved way of deriving the black body spectrum, Einstein suggested – and this was heresy – that the light itself should be treated as if it were actually made of little bits, not waves, each of which carries a *quantum* of energy $E = hf$ corresponding to its color (Einstein, 1905c). This idea has observable consequences, notably for the *photoelectric effect* in which electrons are ejected from metallic surfaces exposed to light.[6] Unfortunately the experimental data at the time were inadequate to confirm Einstein's proposal. More important for the immediate progress of quantum theory was Einstein's subsequent explanation of "missing" contributions to the specific heats of solids (Einstein, 1906). The idea is that each atom in a heated solid tends to vibrate with a definite frequency f as if it were attached with springs to its neighbors. At low temperatures, Einstein suggested, the average thermal energy per atom falls below the Planck threshold hf needed to excite some modes of vibration, and they will not contribute to the heat content of the solid. Thus the specific heat will be anomalously small at low temperatures (Figure 3.2). In this case data were available to confirm the theory, and other scientists began to pay serious attention to quantum ideas. In each application of the quantum concept, a vibratory or rotational motion – one that possesses a definite frequency of recurrence f – does not appear to be excited until the energy provided to it increases

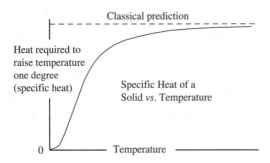

FIGURE 3.2 Heating a solid material causes the atoms to vibrate with various frequencies. As the temperature increases, the thermal energy can excite vibrations at higher and higher frequencies, causing the specific heat to rise. Classical theory predicts that *all* the vibrations would be excited at *any* temperature and the curve would be a straight line as shown.

beyond a certain point. It gradually became clear that whenever energy is exchanged with such a system, it comes in quanta of energy $E = hf$. The simplest picture was the one Einstein advocated, that it is not only the *exchange* of energy with such systems, but the actual energies of the systems themselves, that are "quantized" (Einstein, 1905c).

Atomic spectra. The quantum principle turned out to be the key to accounting for the spectrum, not only of thermal light, but also of radiation emitted by atoms excited by other means – for example, by passing an electric current through a gas, as in a neon sign. Light of this kind is not in thermal equilibrium with its surroundings. When you view it through a prism or some similar device, most of the rainbow of colors is dark, with only a few lit up. The resulting spectrum of bright *lines* is a characteristic signature for each substance. Chemists use these *line spectra* to distinguish among different elements or compounds. In 1913 Niels Bohr (1885–1962) discovered how to use the quantum principle in combination with Newton's laws of motion to "explain" the spectra of atoms (Bohr, 1913).[7] His work launched an entire industry of atomic mechanics that sought to account for the many new phenomena that were disclosed in laboratories around the world by ever more powerful instrumentation.

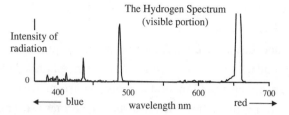

FIGURE 3.3 Intensity of light versus wavelength in the visible portion of the hydrogen spectrum. The intensity of the brightest line at 656.7 nm (red) exceeds the scale of the measuring device. The remaining lines visible here are light blue (486.1 nm), deep blue (434.5 nm), and violet (410.2 nm). Other lines exist in the invisible infrared and ultraviolet parts of the spectrum. Wavelength is inversely proportional to frequency, so in this spectrum red is to the right. Compare with the smooth thermal spectrum of Figure 3.1. (1 nm equals one billionth of a meter).

Bohr pictured the atom as schoolchildren do today: a tiny but massive nucleus with a positive electric charge that binds the much lighter negatively charged electrons in orbits similar to those of the planets around the Sun. Most orbits, however, are not "allowed" in Bohr's theory, only those with certain energies. Bohr chose the allowed orbits so the loss in an electron's energy as it "jumps" from orbit b to orbit a (write it as $E_b - E_a$) appears (by a then unknown process) as a quantum of light whose frequency f is given by Planck's relation $E_b - E_a = hf$. Thus the discreteness of the spectrum is explained by the discreteness in energy of the allowed orbits. That the whole scheme could be made to work for the spectrum of even the simplest atom – hydrogen, with only one electron – was astounding because Bohr's rules for the allowed orbits had no explanation. Like Planck's formula, Bohr's model was impossible to justify on the basis of any known physical principles (Who "allows" the orbits? How does the electron decide when to "jump"?) But it accounted accurately for the spectrum of hydrogen, and it attracted much attention, including eventually a Nobel Prize for Bohr in 1922.

3.4 THE MATTER WAVE OF DE BROGLIE
AND SCHRÖDINGER

Nearly two decades after Einstein's first work on quantum proper-ties of light (and an intervening world war), Prince Louis de Broglie (1892–1987) realized that Einstein's energy–mass relation $E = mc^2$ implied a striking result for particles that have mass. If energy is associated with a frequency according to Planck, *then every mas-sive particle has an associated frequency* $f = E/h = mc^2/h$. Moreover, as the particle moves, the process vibrating with this frequency, whatever it is, will trace out a wave in space with a wavelength that is related to the frequency of the process and the speed of the particle. That is, if Planck's relation is taken seriously, then *every massive particle has a wave process associated with it.* De Broglie showed in 1923 that the wavelength L of this process must be inversely proportional to the particle's *momentum* (mass m times velocity v for slow particles, usually represented by the letter p), the relation being simply $L = h/p$, where h is Planck's constant.[8] Shortly after, wave-like properties of electrons were actually observed with the predicted wavelength. De Broglie's insight became the key that quickly unlocked the remainder of the theory.[9]

So now physicists were aware of phenomena in which light, the quintessential wave phenomenon, behaves like a particle (Einstein's "bits" of light), and other phenomena in which electrons – apparently localized pieces of atoms – exhibit wave-like properties. Thus appeared the famous *wave–particle duality.* Let us focus for the moment on the wave associated with electrons. At the core of the duality issue is the question *What is it that is waving, and what role does it play in nature or in physical theory?*

Early in 1926 Erwin Schrödinger (1887–1961) carried through de Broglie's idea by postulating the existence of "something that waves" underlying all matter. For electrons, the something would be a dynam-ical *electron field* analogous to Maxwell's electromagnetic field and often written $\psi(x, t)$, where ψ measures the "something" at location x and time t. Schrödinger looked for equations like Maxwell's that would have the properties de Broglie had suggested.[10] He was guided by a principle – the *correspondence principle* developed by Bohr and exploited by the atom mechanics – namely that when the objects under examination become large, their laws of motion must approach the usual macroscopic laws of Newton. There is a fascinating hidden wave-like aspect of Newtonian mechanics that has a long history and a beautiful mathematical expression which, unfortunately, I have no time to recount here. We recognize it with hindsight in early nineteenth-century work of the great Irish mathematician William Rowan Hamilton (1805–65), and it came to a climax in Schrödinger's discovery of a wave equation for matter.[11] The solutions of this equation produced, among other things, all the results obtained up to that time using Planck's relation in an *ad hoc* way, including Bohr's results for the hydrogen spectrum. But this theory immediately ran into trouble.

3.5 MEANWHILE, BACK IN COPENHAGEN ...

During the 20 years prior to de Broglie's insight, disrupted though they were by World War I, physicists had managed to explain many new phenomena using Planck's relation and the ideas of Einstein and Bohr. No consistent or complete theory yet existed, but certain recipes seemed to work, and gradually a point of view emerged. The intellectual leader of this process was Niels Bohr, and he thought more deeply about the requirements of a microscopic theory of Nature than anyone else, including Einstein. Others contributed substantially to the theoretical structure, but Bohr guided its interpretation. Just before Schrödinger prepared his papers on his beautiful wave equation, a group of colleagues within Bohr's circle, among whom Werner Heisenberg (1901–76) and H. A. Kramers (1894–1952) were at the time pre-eminent, put the finishing touches on a theory Heisenberg had invented in the summer of 1925. Called *matrix mechanics*, it denied the possibility of a description such as Schrödinger envisioned for his wave.[12] It seemed to them that Schrödinger had not been paying attention, and that his theory actually represented a step backward.

The problem is that electrons seem to be easily detectable as particles localized in space and time. Where are they located within Schrödinger's distributed "electron wave"? Furthermore, in Schrödinger's theory an atom with two electrons must have a waving field that depends on the locations of *both* particles, as in $\psi(x_1, x_2, t)$. This is not at all like Maxwell's waves where for each point in space there is a unique value of the electromagnetic field that does not depend upon any other point in space. Maxwell's wave depends on three space coordinates (one each for height, width, and depth with respect to a reference frame); Schrödinger's wave for two electrons depends on a total of *six* coordinates. For three electrons, it would have nine coordinate variables, and so on. This is a radically new way of describing matter that posed a huge challenge to the interpretation of the theory.

Bohr invited Schrödinger to Copenhagen late in 1926 to discuss the meaning of his new theory. During the intense exchanges Schrödinger fell ill and took to bed in Bohr's house, but Bohr followed him to his room, pressing him relentlessly about the interpretation of the waves, and going through the arguments that had led the Copenhagen group to their own conclusions.[13] Clearly the wave could not be a real, energy-carrying field like Maxwell's. But the experiments showed electrons diffracting around obstructions and acting in other respects like waves. So what is waving? And what is its role in Nature?

3.6 MAX BORN'S STATISTICAL INTERPRETATION

It fell to Max Born (1882–1970) in 1927 to state clearly what today is the generally accepted interpretation of Schrödinger's wave.[14] So as not to introduce too much extraneous material, I am not going to recount how this interpretation developed from the various experiments and ideas that emerged during the preceding quarter century. I will simply frame in language as clearly as I can the conceptual picture of the

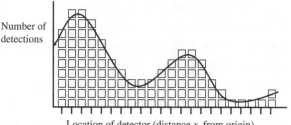

Number of detections

Location of detector (distance x from origin).

FIGURE 3.4 How the Schrödinger wave $\psi(x, t)$ is related to real-world observations. A detector is placed at distance x in an apparatus that prepares an electron repeatedly in the same state. Each repetition is a "trial," and each trial in which a detection occurs at time t is recorded as a box in the column above x in the figure. After a fixed number of trials the detector is shifted to a new position, and the process repeated. As the number of trials at each x is increased, the shape of the tops of the columns approaches that of the squared magnitude of the Schrödinger wave, which is conventionally written $|\psi(x, t)|^2$ (dark curve).

microscopic world called quantum theory. I will not yet attempt to reveal the whole thing – just the part that appeared first and applies to the simplest situation: a universe with just one electron. Call this Step 1 on the way to the whole picture. (There are only two steps.)

We now understand that the Schrödinger electron field is not real in the sense that Maxwell's electromagnetic field is (in pre-quantum physics). It does not correspond to a force, and it does not directly carry energy or momentum. Rather its *intensity* (squared magnitude) at each point of space, times the volume of a small region of space surrounding the point, is the *probability of observing an electron in that region* (Figure 3.4). Probability here has the usual practical meaning – a number equal to the fraction of repeated trials that give a specified result when the number of trials is very large. The probability of rolling five with a pair of honest dice, for example, equals one ninth because out of the 36 ways the dice could fall, four ways display five dots, and $4/36 = 1/9$. "Honest" means none of the 36 ways is favored. Probabilities are pure numbers between zero and one. They do not *measure* anything, and therefore do not come in units like centimeters or pounds. *Stating a probability means you have in mind what result you are looking for,* in this case detecting an electron at a definite time and place. Matching a probability to experience implies a large number of identical trials with random outcomes. The probability concept is normally used when the outcome of any particular trial cannot be predicted accurately, but where a large number of trials exhibits statistical features, such as the *mean value,* that might be predictable.

Because it represents *probability,* the Schrödinger field can be related to statistical properties of a large number of observations. It usually does not give full information about a single observation. Keep the dice example in mind. Knowing the probability of a five-score, or of all the other possible scores, does not tell what the next roll will bring. And it may take many trials before the average of your results begins to approach the theoretical one-ninth. In many – but by no means all – applications, quantum theory is useful because phenomena of interest to humans usually involve huge numbers of individual microscopic events. The electron beam that draws pictures on the face of a (now obsolete) cathode ray television picture tube, for example, carries about a thousandth of an amp of current, but this equals about six million billion (6×10^{15}) electrons per second. If the beam were defocused slightly, and then allowed to rest at one place on the screen, it would be bright in the center and fading toward the edges in a characteristic pattern. The squared amplitude of Schrödinger's field for any one electron in the beam describes this pattern. (In this imaginary experiment I am ignoring thermal and other phenomena that might smear the beam much more than quantum effects.) Each electron that arrives at the screen makes a tiny flash somewhere within the pattern. If the current were turned way down to one electron per second, you would see flashes at random places. Only in a long time exposure would you see the pattern emerge that is predicted by the Schrödinger field. We do not have a physical theory that allows us to predict where or when the individual flashes occur. We can only predict the overall pattern. To be precise, the number of times a flash appears in a little dot in one second should approach the intensity of the Schrödinger wave at that dot times the area of the dot times the number of electrons emitted toward the screen in that second (which is the number of trials in the dice analogy).

Schrödinger's approach to quantum mechanics turned out to be more visualizable, but less suggestive of the full generality of the theory than Heisenberg's. As Schrödinger himself quickly demonstrated, the two approaches are mathematically equivalent. Schrödinger's emphasis on waves, however, is psychologically misleading.

The probability aspect of the wave function is more general than the "wave-like" aspect. I wish we could replace the word "wave" with something else, like "information" or even "Born," as in "Born function," but this is not the worst etymological misfortune in quantum theory, and I simply warn you here once and for all that the word "wave" is misleading. Schrödinger's original wave was for an electron, and the set of events for which it supplied probabilities were detection events in a hypothetical electron detector placed at different locations, much as Hertz mapped out electromagnetic waves by carrying a circuit from place to place (see Note 22 in Chapter 2). "Wave functions" exist, however, for *any* detection event, not just for the locations of electrons. The "functional" aspect of wave functions relates any set of possible detection events to a corresponding set of probabilities. An example is the wave function for Schrödinger's famous cat, described below. Deeper examples appear throughout the remainder of this book, by the end of which you will, I hope, emerge with a clearer idea of the concept of the wave function.

3.7 THE QUANTUM MICROSCOPIC WORLD VIEW: STEP I

This all sounds innocent enough, but what does it mean for our conception of Nature? Everything outside ourselves (ourselves too, but let's keep it simple) is represented in quantum theory by a wave function that gives the probability that an incident of sensation will lead to a certain specified result. Everything we know about the world consists of discrete incidents of sensation, flashes of nervous impulse triggered by a messenger carrying energy and momentum from the outside world. We experience these incidents as a continuum because our crude instruments of perception (eyes, ears, ...) smooth over the myriad individual detection events. Each such event – I like to call them *clicks*; Bohr called them *registrations* – has a definite outcome. There is no uncertainty regarding the world we actually see. It is a network of sensory clicks. Each conscious being possesses a unique memory of a world of clicks (so to speak). When we compare our very definite world with another's, we are not surprised that their pattern of clicks is not exactly the same as ours, but statistically speaking they are sufficiently close. In the world of large things, the deviations in the pattern of sensory events between two observers is small. We can speak to one another of what we experience, and expect to be understood.

In the world of the small, however, human perception lacks the necessary resolution. Observers must objectify experience by capturing clicks in some macroscopic external medium, such as a photographic emulsion, that any observer can then peruse. Bohr's word *registration* conveys the flavor. The objectifiable experience is one in which a microscopic event has been magnified and registered irreversibly in apparatus resembling a mousetrap. *All* detectors have this property, including our own senses. A tiny transfer of energy by a message from the system we are observing provokes an irreversible avalanche of events in the detector that creates a sort of phenomenal lump ponderous enough to resist destruction by further gentle probing. What is real to you and me is these lumps, these registrations of microscopic events that comprise the macroscopic world. *Any* phenomenon large enough to measure (as with a ruler) serves as a detector of some prior interaction with the microscopic domain. In this way the frustrating elusiveness of objects too small to pin down is translated into a series of well-defined features of a universe that anyone can examine.

Reality, as I understand the word, is ultimately a social phenomenon. If other witnesses cannot see it for themselves – or reproduce the experience of seeing it, or imagine a way that it might have been reproduced or seen by others – then it is not physically real. I admit this criterion is vague and needs to be refined and distinguished from other possible uses of the word "real." Bohr would say more definitely that if the phenomenon is not *registered* it is not real. Only registered phenomena can be shared. Clearly "reality" is a term that makes most sense for macroscopic things. Perhaps we should call it "experiential reality" or even "existential reality." Quantum theory causes headaches because we cannot help using our macroscopic language – whose underlying world view may even be hardwired into our brains – to speculate about the microscopic world. The macroscopic laws of motion, the laws of Newton and Einstein, track the movement of "big atoms" with well-defined positions at each instant. The macroscopic picture of the world is one of a tapestry of continuous big atom world-lines in space-time. The microscopic picture, however – well, *there is no microscopic picture*. There is a discrete set of registrations, each of which is a marker that may together with other markers suggest a track approximating the world-line of a macroscopically observable object in space-time. But there are no world-lines *in our experience* corresponding to the microscopic elements themselves prior to registration. The tapestry breaks up into a pointillist scattering of isolated clicks.[15]

So much for experience. What about theory? Raw experience is chaotic even in the macroscopic world. Theory helps us organize this chaos by proposing models whose components simulate the things we actually see. If the theories work, that is, if they appear to simulate accurately the pattern of observable events, then we are inclined to attribute a kind of reality to their components. We might call these models *theoretical reality*, which sounds like an oxymoron. This is the kind of reality, however, most relevant to human affairs. It is this image of reality that persists in conscious memory and motivates future actions. Short of inadvertent knee-jerks, or a more or less conscious surrender to impulse, we rarely act upon raw unfiltered experience, but rather upon images based on past experiences that we carry in our minds. These are the images that give meaning to experience. They are the visions that suggest possible consequences of actions, and they have an extraordinary psychological power. Only these remembered – *registered* – images are continuous and permanent. The actual experience of physical reality is fragmentary and fleeting. So we attribute persistence or trajectories or world-lines to all visible objects despite the hopelessness of observing them directly at every instant. We think of the world-lines as real because the theory that contains them – the possibly hardwired model in our brains – works spectacularly well for big things. One of quantum theory's striking features is that it does *not* have any component resembling trajectories or world-lines for microscopic objects. That is, *the quantum world view does not fill in with theoretical entities all the parts of Nature that seem to be missing from our direct experience.*

3.8 SCHRÖDINGER'S CAT

Quantum theory, in the Copenhagen view, does not tell all we would like to know about things. It does not attempt to describe "things" at all, only their potential impact on our senses (or on any other registration device). This is a kind of theory radically different from the classical theories of Newton or Maxwell, which are similar to mechanical models whose various parts accurately simulate "objectively real" entities in Nature. Yet quantum theory does have some of the flavor of a mechanical model. Among its features are a well-defined wave function and an equation of motion. It is tempting to think of the wave function as in some way representing "something real," and the temptation is greatest when the "things" are small and simple.

But real things can be large and complicated. Schrödinger envisioned a wave function for a cat to emphasize how absurd it is to imagine wave functions are "real things." The wave function does not have to look at all like a wave. Its key feature is a list of probabilities for registrations corresponding to a set of well-defined events – one probability for each event. For the cat the events are determinations that the animal is dead or alive. Schrödinger imagined inserting a cat in a box along with a flask of poisonous vapor linked to a device that would smash the flask when a detector clicked in response to the decay of a radioactive nucleus – all loaded into the box together. Radioactive emission occurs randomly with a characteristic average time between decays – the *half-life*. After the lid is closed, you wait one half-life. At that time, quantum theory implies a wave function that gives the cat a 50% chance of being observed alive when you open the box, and a 50% chance of being dead. Somehow the wave function must accommodate *both* possibilities, but the cat is obviously *either* dead *or* alive before you open the box. And before you open the box, according to quantum mechanics the wave function contains all the knowledge you can have of the cat. Therefore the wave function cannot possibly be an adequate description of the "real" cat.

Schrödinger did not intend this example to be a paradox, but wanted specifically to make the point that the wave function cannot be a model of reality.[16] Nor did he suggest that the observer who opens the door decides the cat's fate. Its fate is sealed as soon as a radioactive emission effects an irreversible consequence in the world – certainly by the time the first detector clicks. We simply do not know what has happened until we open the box. If we want to reassure ourselves that our action did not kill the cat, then we can perform an autopsy to determine the instant of demise.

This example is useful for exposing the rather awkward way physicists have forced quantum theory to fit the conditions of actual experiments. From the time t_1 when the box is closed until the time t_2 when it is opened, quantum theory represents the box and its contents with a complicated wave function that obeys a set of Schrödinger equations – one equation, roughly speaking, for each "particle" comprising the apparatus. The observer provides the initial conditions at t_1 from which in principle the evolving wave function can be calculated up until t_2. But at the time of opening, something happens that is not described by any process either within quantum theory or in classical theory. There is a discontinuity at t_2 where the computed wave function becomes irrelevant and the observer becomes aware of the actual state of affairs, in unambiguous macroscopic language, as Nature has arranged them. Which macroscopic possibility Nature has selected depends partially, but not entirely, on what the observer has chosen to observe. I will say more about this observer-dependence in Chapters 4 and 5, but here I simply want to point out that the uneasiness expressed by Schrödinger about the whole conceptual picture is well-founded. That leap at t_2 when the observer decides to stop calculating and just look to see what is going on is euphemized by hypothesizing a nonvisualizable time transformation called "reduction of the wave function." That physicists become tongue-tied at this point is embarrassing, but not surprising.

We *do* understand why the discontinuity is there: *any* theory based ultimately on a propagating probability function must stop short of predicting what actually transpires. You cannot have the probabilistic form of quantum theory without that final step, and that step is necessarily incomprehensible within the smoothly evolving domain of Schrödinger's equation.

If we do not want to accept this unexplainable leap in our theory, we have two choices. We can look for a new theory that is not founded on probabilities, which no one has succeeded in doing satisfactorily (but I admire the spirit of those who persist). Or we can change the concept of "reality" simply to avoid the leap altogether. In the *many worlds interpretation* of Hugh Everett, "reality" includes all those alternative possibilities for which the wave function predicts probabilities (Everett, 1957). In that case Nature does not have to choose among the possible outcomes – they are all part of a fantastically extended "reality." "Many worlds" refers to the infinitely multiple universes this model requires, each of which exists in parallel with, but mutually incommunicable with, all the others. This interpretation, while logically conceivable, seems to me an extravagant violation of Occam's razor. The Copenhagen approach is to accept the leap, more or less in the spirit of Newton's "I frame no hypotheses," and move on.

Quite a different awkwardness is the apparent ambiguity in the time t_2 of the leap. The cat itself is present inside the box to sniff the poisonous vapor and hear the hammer fall, or possibly even to see a spark in the radiation detector, each of which perceptions essentially short-circuits the chain of events that holds the external observer in suspense. We could locate t_2 at any of these events, and the ultimate outcome would not be changed, only how we portray it to ourselves in theory. The boundary between the two regimes of evolution – "quantum theoretical" and "unexplained natural phenomenon" – is movable.[17] And yet it cannot be too close to the starting point t_1. Something complicated happens when a detector clicks. A "simple" system like a single decaying nucleus interacts with the large and complex system of the detector. Very quickly, the number of coordinates ("degrees of freedom") required to describe the parts set in motion by the radiative decay becomes huge, and the wave function becomes correspondingly more complicated. Eventually, however, the wave function "settles down" to describe a set of mutually independent possible outcomes (e.g., "click" and "no click").

Bohr grasped intuitively the mouse-trapping that converts the complicated evolving wave function irreversibly into a set of probabilities for well-defined macroscopic registrations. Much of the century passed, however, before physicists developed a theoretical account of this process. It is complicated by the fact that irreversibility entails the disturbance of many pieces, as in the dissipation of energy in friction, or the dispersal of a drop of ink in water. The quantum version of the process is called *decoherence*, but the image of a mousetrap will serve our purpose.[18] The duration of the decoherence process essentially determines when it is "safe to look" for a stable macroscopic indicator of what Nature hath wrought. If you press t_2 too closely against t_1, then you inevitably participate in processes for which only quantum theory has been found to be valid. You become part of the "quantum system," which leads to the possibility of an entirely new technology in which the deliberate intervention in the quantum machinery leads to control of phenomena unheard of in the classical world. More about this in Chapter 5.

Unfortunately, from the philosophical point of view, the unerring success of the "naive" Copenhagen approach undermines passion for finding something more in line with our prejudices about what a theory of Nature should accomplish. This fosters a habit of what J. S. Bell famously called "unspeakability" in quantum mechanics (Bell, 2004, paper 18). *Any* description of Nature, however, based ultimately on a probability function necessarily includes the leap described above. It is an intrinsically unresolvable feature of quantum theory. And it is not, if you can believe it, the weirdest feature of quantum theory. See Chapter 5.

3.9 WAVES VERSUS PARTICLES

The early literature on quantum theory, which is still widely read by scientists as well as by philosophers and historians, is an archaeological midden in which the precursors of our current understanding are mixed with misleading and irrelevant debris. Much confusion has been created by authors who have tried to reconcile this old language with our modern understanding, especially about the issue of waves versus particles. So let me state clearly: One of the fundamental concepts of quantum theory is a wave, or more precisely, a *function* that for an important class of physical systems assigns a number to each point in space, but it is a *probability function* that does not signify a measurable quantity like energy or momentum, and can be considered "real" only insofar as the theory that contains it is successful. On the other hand, at the fundamental level, the theory does *not* contain "particles" even as theoretical entities. What it does contain is *registrations*, and these do not appear as components of the theory, but in the interpretive statements that link the theory with experience (Born's interpretation).

Quantum theory provides for each small region in space and time a number equal to the probability of registering an event there. It does not provide a visualizable model for the event. In particular, it does not suggest that the event is perpetrated by a spatially localized little object that persists continuously in time, which is what I take the word "particle" to imply. On the other hand, quantum theory does permit us to observe objects that are small by human standards and that do persist in time, so long as we do not attempt to fix their dimensions too narrowly: marbles, dust motes, molecules, even – most of the time – single atoms and nuclei. This must be possible because quantum theory must be consistent with the very successful macroscopic theory (Newton/Einstein or Maxwell) in its large-scale domain of validity.

Other physicists might at first disagree with the idea that quantum theory has no particles. When pressed, however, they will admit that "Well, we don't use the word 'particle' in an ordinary sense," and then go off on a long explanation of what physicists really mean by "particle" (see Chapter 7).[19] I am saying it were better if we just did not use the word "particle" at all for whatever causes microscopic detection events, because what everyone other than physicists means by the word is perfectly clear. And it is perfectly clear that the ordinary particle concept is not part of the microscopic theory. Most authors of quantum mechanics texts apologize for using the word "particle" for the parts of microscopic Nature, but then use it anyway. The result has been widespread confusion, even among scientists. The oxymoron "quantum particle" is a barrier that inhibits public access to the marvelous world beneath our senses. It should be abandoned.

3.10 ABOUT WAVES

De Broglie's famous discovery of a link between motion (momentum) and wavelength has consequences the appreciation of which requires some wave-lore. *Waves* in Nature are moving displacements of something you could measure, like the depth of water in an ocean wave, the pressure in a sound wave, or the electric force exerted by a radio wave. In most of these examples the displacement repeats itself after a certain time (the wave *period*) and distance (the *wavelength*). These are *periodic waves*, of which the *sine-wave* is the notorious exemplar. The word *sine* comes (through a tortuous history of mistranslations) from a Sanskrit phrase meaning *half-chord*, referring to half the line segment on a slice of a circle. Plotting the length of the half-chord versus the angle it subtends at the center produces the famous sine-wave (Figure 3.5).

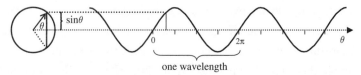

one wavelength

FIGURE 3.5 *Origin of the "sine-wave."* The angle θ (Greek *theta*) is half the angle of the wedge whose chord is shown as a bold vertical line in the circle at left. When the radius of the circle is "one," half the chord is called the *sine* (abbreviated "sin") of the angle, and its value is graphed at right as θ advances counter-clockwise. If θ is measured in units of *radians* the wavelength of the sine-wave equals the circumference of the circle, or 2π times its radius (180° equals π or 3.14159... radians). As θ increases, the pointer sweeps around the circle like the hand of a (backward-running) clock.

Unless the waving medium is confined so it sloshes back and forth, naturally occurring waves always tend to travel outward from a source, and waves of different wavelengths tend to travel with different velocities. It should be clear that a periodic wave moves one wavelength in one period. The successive peaks wash over a stationary observer at the rate of one per period, which is called the *frequency* of the wave. The wave velocity v therefore equals the product of wavelength and frequency, $v = Lf$, so the frequency is inversely proportional to wavelength, $f = v/L$. Light waves with different frequencies all have the same velocity c in vacuum, and very nearly the same in air. Optical spectra can be graphed as a function of either frequency (as in Figure 3.1) or wavelength (Figure 3.3) of their light waves.

Periodic waves can have any shape, including "sawtooth" or "square," but sine-waves are important because of their simple mathematical properties. All the other kinds of waves, even nonperiodic ones (like a tsunami), can be regarded as made up by adding, or *superposing*, sine-waves with different wavelengths. Each component wave typically will have a different height (*amplitude*) and peak location (*phase*). When they are summed the parts of the sine-waves below the zero amplitude line ("negative phases" where the arrow in Figure 3.5 points downward) subtract from the rest, so there is a complicated

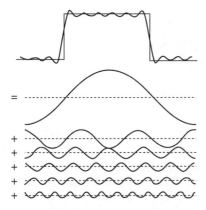

FIGURE 3.6 Decomposition of a square shape into a sum of sine-waves. The wiggly pattern in the rectangle shape above is the sum of the six sine-waves below it. Adding more sine-waves improves the match between the rectangle and the sum.

pattern of cancellation and reinforcement. Using the mathematical properties of sine-waves and a little calculus you can figure out exactly what the amplitudes and phases of the superposed sine-waves must be to make up any wave shape. The details are not important for under-standing quantum mechanics, only the concept. Figure 3.6 shows a few of the sine waves in the sum that makes one period of a square-shaped wave.

Thus every wave shape is a unique superposition of sinusoidal waves, each with its own wavelength, amplitude, and phase.[20] The percentages or relative amplitudes of each component can be displayed as the wave's *spectrum* – amplitude vs. wavelength or frequency – which is the signature of the wave phenomenon. In the example of Figure 3.6 the amplitudes diminish as the wavelength gets smaller, and successive components have opposite phases. The superposition is obviously dominated by the wave whose peak is about the same width as the square. Many home music systems have an illuminated display of the frequency spectrum of the sound signal going to the speakers. The display consists of a row of illuminated bars, one for each frequency band, from bass to treble, that grow or shrink as the music has more or less of that frequency component in it. If there were many more bars, the display would give an alternative "spectral pic-ture" of the music signal.

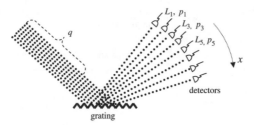

FIGURE 3.7 How to measure the momentum spectrum of a Schrödinger wave. The grating, which acts like a prism, could be the surface of a crystal whose regularly spaced rows of atoms diffract a rainbow-like spectrum of wavelengths, L_1, L_2, L_3, ... each of which corresponds to a momentum *parallel* to the original beam direction. The pattern of detector clicks might resemble Figure 3.4, where the distance x is measured *perpendicular* to the beam direction. Here q is the location *parallel* to the beam. An apparatus to register clicks at distance q would obviously obstruct the measurement of p.

It is a fact of the utmost importance that wave shapes can be defined *either* by their shape in space, *or* by the shape of their wavelength spectrum, which you might regard as a picture of the wave in "wavelength space." The two descriptions or representations are *complementary*, a word introduced for a similar purpose by Bohr. The spectral picture of a single sharp peak in ordinary space is a broad band in wavelength space because many short wavelength waves have to be superposed to cancel each other outside the peak. The spectral picture of an undulating wave extending over a broad spatial region is a single narrow peak centered at the wavelength of the undulation (see Figures 4.5a,b). Sharpness in one representation *necessarily* corresponds to broadness in the complementary representation.

The momentum spectrum. Schrödinger fields heave in their own inscrutable medium, and carry forward a spectrum of elementary waves. According to Born's interpretation, *the amplitude (squared) of each elementary wave is the probability that a momentum detector will click in the presence of the field when its momentum is set to the value corresponding to that wave's wavelength* (Figure 3.7). De Broglie's relation links the elementary wavelength and the detector's momentum setting in the simple formula $L = h/p$, i.e., Planck's constant divided by the momentum (mass times velocity for slow objects). A wave function with a single peak describes a phenomenon with a well-defined position. But many sine-waves of different wavelengths must be superposed to make up such a peak, and clicks will occur over a range of settings of a momentum detector looking at the same wave function. The momentum of the underlying quantum object is therefore not unique. It might be said to possess a *spread* of momenta. An object, on the other hand, whose Schrödinger wave rolls sinuously along with regularly repeated peaks and troughs has a well-defined peak-to-peak wavelength, and its momentum is correspondingly well-defined. Since the amplitude of such a wave is constant throughout its course, a detector placed anywhere has the same probability of clicking. Therefore the position of the underlying phenomenon is not well-defined.

3.11 THE "UNCERTAINTY PRINCIPLE"

The Schrödinger field pattern in *position space* determines *where* a detection event is likely to be found, and its pattern in *wavelength space* determines the *momentum* we associate with the object causing the event. If the events are localized in a small region, the wave pattern will be localized but consequently it will contain many elementary waves – *its momentum will not be well-defined.* Conversely, if the momentum detector clicks only for a narrow range of momentum values, the wavelength is well-defined, and the wave pattern must extend over many cycles – *its location in space is not well-defined.* You can have waves with well-defined position *or* well-defined momentum, but not both at once. This is the true meaning of the *uncertainty relation* first proposed in 1927 by Heisenberg.[21] To be precise, the relation states that among many measurements on a large set of systems, each prepared identically, the range of momentum values times the range of position values you can expect to find is greater than or equal to Planck's constant times a certain number. The number depends on how you define the ranges or spreads, commonly written Δq and Δp, where q and p stand for position and momentum. For the most common statistical definition (the *standard deviation)* the constant is $1/4\pi$. Thus the algebraic formula: $\Delta q \Delta p \geq h/4\pi$. A similar important relation exists between the range of measured energy values ΔE and the interval of time Δt available to make the measurement: $\Delta E \Delta t \sim h$.[22]

The "Heisenberg uncertainty relation" emerged in an atmosphere of confusion from which it has never quite escaped. Much of the fault lies with Heisenberg himself who was not content with setting forth the bare idea through mathematical analysis – he also tried to make the result more comprehensible with suggestive physical arguments. His examples encouraged the notion that the uncertainty has its origin in the inevitable disturbance caused by the measurement process (which is not inherently a quantum concept). Bohr objected to these explanatory efforts, convinced that the matter was deeper than Heisenberg made it out to be. As I see it, most problems of interpretation are resolved by the simple fact that the microscopic theory does not refer to any physical waves *or* particles. It refers to well-defined detectors and unambiguous events of detection. Accounts that ascribe position or momentum to waves or particles apply macroscopic language inappropriately to microscopic Nature. You can set a detector to register an event with well-defined momentum, or you can set it to record an event with well-defined position. That does not entitle you to say that the event is caused by a "wave" or by a "particle." Chapter 4 gives more insight into the uncertainty relation.

Quantum pressure. One important physical effect implied by the uncertainty relation is a kind of quantum pressure. A detector within a tiny box that contains electron stuff will trigger only if the Schrödinger field contains very short wavelength components (so they can fit inside the box). According to de Broglie's relation, such a field describes an object with large momentum components, which must bounce off the walls of the confining box, creating a pressure against which the walls must be braced. In other words, to confine electron stuff, whatever its physical nature, necessarily requires a force that must increase as the confined volume is made smaller. This is the origin of the bulk compressibility of matter. Remarkably, the size of an atom (i.e., of its "electron cloud") can be estimated approximately by balancing this outward pressure from the uncertainty relation against the inwardly directed electric force.

"Quantum pressure" is one of Nature's most important forces, and yet it does not appear in the standard catalogue of four: gravity, electromagnetism, weak and strong nuclear interactions. It emerges Zen-like out of the simplest facts of quantum theory, and should be recognized more explicitly for what it is: the force that balances electromagnetism to define the world of human scale in the vastness of the universe. There is a lesson here in the surprising obliquity with which theory "explains" phenomena. Facts of the most urgent human consequence sometimes appear as mere side-effects of deep properties of theoretical entities.

Ground state energy. Another phenomenon required by the uncertainty relation occurs in all oscillating systems. Think of a pendulum swinging through a small arc. Its total energy is the sum of a "kinetic" part that depends on momentum and a "potential" part that depends on position (displacement from vertical). The least energy state, zero, occurs when the pendulum is hanging straight down and not moving at all. But in that state both the position (down) and momentum (zero) are simultaneously well-defined, which is forbidden by the uncertainty principle. Thus the detected least energy of a pendulum, and indeed of any oscillator, cannot be zero! Quantum oscillators always register a small zero-point motion and a corresponding *ground-state energy*.

Maxwell's equations imply that the energy of electromagnetic radiation includes a magnetic and an electric component (see Note 2 in Chapter 3). When the fields are treated quantum mechanically, the magnetic field behaves somewhat like a position and the electric field like a momentum so the two obey an uncertainty relation. Therefore, the *fields in an electromagnetic wave will never be found to vanish simultaneously*, and for every frequency there will be a nonvanishing zero-point electromagnetic motion. Certain effects of these zero-point fields, or *vacuum fluctuations* are observable, but the overall picture of a nonvanishing vacuum energy for each of the infinite number of possible electromagnetic frequencies is puzzling. The sum of all these small energies would diverge to create an indefinitely large, all-pervasive energy density throughout space. This would lead to gravitational effects incompatible with astronomical observations.[23] The observed effects of vacuum fluctuations can nevertheless be calculated from the theory because they involve only *differences* in ground-state energies in which the embarrassing infinities cancel.

3.12 AMPLITUDES AND PHASES

We are closing in on an aspect of the quantum world that is important for our story. The primary wave-like object in the theory, the Schrödinger wave function, has an odd and crucial property, unprecedented in science. What the theory actually produces is *two* numbers at each space-time point: an *amplitude* (call it *a*) whose squared value is the probability of a registration, and a *phase* (call it θ). The phase is usually expressed as an angle that describes a mathematical process like the rotating pointer in Figure 3.5 that returns to its starting value after each multiple of 360° or 2π radians. Amplitude and phase appear together in a particular combination that can be represented by a *complex number* about which I will say more in the next chapter. Schrödinger himself was not happy about the complexity of his wave, but the logical requirements of the theory cannot be met without it.[24]

You can see already in Figure 3.5 that the sine-wave pictured there is somehow less fundamental than the pointer with its length (amplitude) and angle (phase) whose motion sweeps out the wave pattern. The sine is just the *projection* of the pointer upon the vertical direction – the shadow, as it were, cast by the pointer on a vertical wall by a light shining from the left. As the pointer turns, this particular projection first peaks at a phase of 90° or π/2 radians. A horizontal projection would lead to a similar wave, but it would have its first peak at 0°. This is called a *cosine-wave*, identical to the sine-wave but 90° "out of phase" with it. Other projections would give different waves that peak at different phases (Figure 3.8). The amplitude–phase

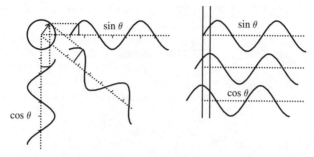

FIGURE 3.8 Each of these three waves has the same wavelength and amplitude, but a different phase. The cosine function differs from the sine function only by a phase shift of 90°.

pair is the fundamental object because it defines the pointer that produces *all* these different waves, depending on which projection you choose. The Schrödinger "wave" is not completely described by any one of the sinusoidal projections. It is the amplitude–phase pair itself, and in a general wave function both members of the pair can vary with time and location. Only the amplitude, however, is linked to the statistics of observed detection events. The phase does not directly affect the statistics and therefore it cannot be directly inferred from them.

Although it cannot be observed directly, the *quantum phase* plays a crucial indirect role in the theory. The Schrödinger field of a moving quantum object can be visualized as an advancing wave whose pattern of crests and troughs can curve in the presence of obstacles, like ocean waves about an island. Where the distorted patterns cross each other (behind the island), the two parts interfere depending on their relative phases. Where both oscillations are at a peak, the wave displacement is enhanced. A trough arriving together with a peak, however, will cancel part or all of the displacement there. It is the *relative* phase, not the absolute phase, of the interfering portions that determines how the two parts will add. Observations do not trace how the different parts of the wave come together and cancel. We only see the final result.

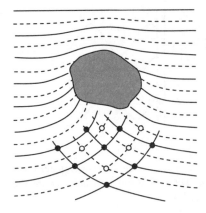

FIGURE 3.9 The slowing of ocean waves in shallow water causes their crests (solid lines) and troughs (dashed lines) to bend around an island. Black dots (white dots) show where crests (troughs) interfere constructively. Destructive interference occurs where troughs and crests coincide.

The quantum phase helps us to understand some otherwise puzzling phenomena. Imagine the Schrödinger wave associated with an electron orbiting the nucleus in Bohr's atomic model. This model appeared in the misty dawn of quantum theory, so it includes the particle-like concept of *orbit*, which nevertheless provides an approximate picture. As the orbit wraps around the nucleus, so does the wave, whose phase is cycling with the period of the de Broglie wavelength of the moving electron. The electron could be detected at any point along the orbit, because the amplitude is more or less constant, while the phase varies from point to point. Now consider the phase of the orbiting wave as it returns to its starting point on the closed path. If it is out of phase with the oscillation at the starting point, then it will *cancel itself* there (outer orbit in Figure 3.10). As it proceeds, it will never get back in phase, and the cancellation will occur around the entire orbit. Consequently, *we have no chance of ever detecting an electron on that orbit*. We will only see electrons on orbits that are *integral multiples* of a de Broglie–Schrödinger wavelength. For these orbits alone will the wave wrap back on itself with the right relative phase for reinforcement rather than destruction (inner orbit in figure). Thus is explained the sequence of "allowed" Bohr orbits – a complete mystery before de Broglie's insight.

A more accurate impression of Schrödinger's three-dimensional electron field for a single electron is conveyed by the surface in

FIGURE 3.10 A visualization of de Broglie waves on two different orbits associated with an electron circling an attracting nucleus. The outer wave "cancels itself out," while the inner one reinforces itself and is thus an allowed *Bohr orbit*. The black dot represents the nucleus which would be invisibly small on this scale.

FIGURE 3.11 A Schrödinger wave "probability surface" for a wave similar to the inner orbit in Figure 3.10, but with half as many peaks and troughs. Positive phase is dark, negative phase light. The axis of circulation is tilted away from the viewer to show the three-dimensional character of the surface. See *The periodic table*, Chapter 6, for other atomic wave functions and more detail about their properties. (Figure generated from the program *"Orbital Viewer,"* version 1.04, by David Manthey, 2004, www.orbitals.com)

Figure 3.11. This figure does not attempt to show the amplitude and phase everywhere, but displays the shape of the probability function (amplitude squared) by its "edge," that is, by the surface where it practically disappears. What you see in the figure is the boundary beyond which the probability of registering an electron event is less than one in a million. Such a figure cannot show the detailed phase, but the regions where the phase is negative (between 180° and 360°) have a lighter shade. Each disc-shaped surface corresponds to a peak (dark) or a trough (light) in the crude wavy orbits of Figure 3.10.

3.13 QUANTUM PHASE AS A NEW "DIMENSION" OF NATURE, AND WEYL'S TRIUMPH

Now we are ready to pick up the thread of the story that began with Bernhard Riemann and the geometry of Nature. Recall that after

Einstein "explained" gravitational force as curvature induced in space-time by matter, Hermann Weyl proposed that changes in the scale of length (the *gauge*) from point to point might "explain" electromagnetism geometrically too. That theory had unphysical consequences. There is no evidence that the scale of sizes in space-time is related to electric or magnetic forces.

According to Schrödinger's theory, however, the phase of the wave function's amplitude–phase pair is advanced or retarded from point to point along a path in the presence of an electromagnetic field. That is why the wave distorts near the charged nucleus of an atom to form its pattern of allowed orbits. The phase, which locates peaks and troughs of the wave function, is retarded more near the nucleus where the electric field is stronger. Thus the outer crests of the wave move faster than the inner ones and the whole pattern wraps around the nucleus. *If you could see how the phase changes from point to point along the various paths, you could tell how strong the electric field is between the points.* So it is worth checking to see if Weyl's idea applied to the scale that relates distance along the path to Schrödinger's *phase* gives anything interesting. It does. To Weyl's delight his entire gauge theory transferred easily to the Schrödinger path-dependent phase. And this time, there were no unphysical consequences.[25] Now I request your patience while I explain Weyl's idea in some detail. You may want to come back to the next few paragraphs from time to time as you read the other chapters, but I urge you to read the following carefully now as it is the key to an important aspect of our current "best" worldview.

Forces and potentials. Before proceeding, I need to explain that the thing that links the quantum phase in Weyl's theory to steps along a path through space is not directly the electric or the magnetic force field, but a related simpler field with a longer name: the *electromagnetic four-vector potential*, or just *vector potential*, with symbol A. *Vectors* are things that, like forces or arrows, possess direction as well as magnitude (see Chapter 4). The electric field E is proportional to the rate of change of this potential in time and space, and the magnetic field B to the amount of spatial "curl" or circulation in the potential field, which is proportional to spatial rates of change in different directions. You can see that A can accommodate some arbitrariness in its definition because it is not the absolute value of A that is important for the fields, but only how it changes in space and time. This property is crucial for linking Maxwell's equation to Schrödinger's. Prior to Weyl's idea, the vector potential appeared to be an auxiliary quantity that was sometimes useful for doing calculations but had no physical reality. Today we recognize it as fundamental in a complete description of electromagnetism.[26]

On mis-measuring quantum phase. Let me restate Weyl's idea in this new context. If you wanted to check the ratio of the quantum phase to the path length for some object described by a wave function, you would need an instrument with a display somewhat like a student protractor that could measure the angle of the Schrödinger phase at any point. (This is a thought-experiment – you cannot measure the phase directly.) The angle you read depends on two things: the actual phase and how you hold the "protractor." If you misalign the instrument as you move it about you will introduce errors in the readings. Only by keeping track of the misalignments can you recover the actual phase. Having made a series of observations, you have two choices. (1) You could just subtract the record of misalignments from the raw data of measured phases, which is equivalent to accurately aligning the protractor in the first place. Then the difference between the corrected phases for two very close points, according to Weyl, equals the electric charge on the object times the vector potential near those points times the distance between the points. Since the vector potential points in a definite direction, you need to multiply by the part of it – the *component* – pointing from one point to the other, as shown in Figure 3.12. So divide the phase difference by the charge and the distance to find the resulting component $A_{\parallel}$ of the vector potential.

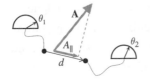

FIGURE 3.12 Relation among quantum phase, location, electric charge, and electromagnetic potential (alternative 1). The quantity $A_\parallel$ is the part of **A** parallel to the line of length d between the two points where the phases θ_1 and θ_2 are (imagined to be) measured with the little instruments that look like protractors. $(\theta_2 - \theta_1)/d = charge \times A_\parallel$.

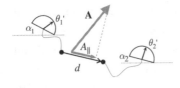

FIGURE 3.13 Same as previous figure, but with the phase "protractors" misaligned (alternative 2). Now the vector potential component needs to be corrected by subtracting a *gauge transformation*. $(\theta'_2 - \theta'_1)/d = charge \times A'_\parallel = charge \times A_\parallel + (\alpha_2 - \alpha_1)/d$.

Alternatively, (2) you could just take the raw phase measurements *without* correcting and use them to calculate the vector potential – call this result A'. The vector potential A' differs from A by a correction proportional to the difference in the phase misalignments at the two points divided by the length between them – a quantity mathematicians call the component of the *gradient* of the misalignments between the points. The second potential is said to differ from the first by a *gauge transformation*. The arbitrary misalignments have *transformed the gauge* (Figure 3.13).

Now comes the deep insight: following Einstein's philosophical guidance, *the laws of Nature should not depend on what arbitrary choice we make for the reference frame of the quantum phase.* That is, neither Schrödinger's equation of motion nor the electric and magnetic fields E and B should depend on the particular list of misalignments in the thought-experiments in the previous paragraphs. The fields, in particular, must not be affected by gauge transformations of the type described above, namely by adding the gradient of an *arbitrary* function (the misalignments at each point) to the vector potential. The arbitrariness in the definition of A mentioned above *is just such as to guarantee this property.*

3.14 ELECTROMAGNETISM "EXPLAINED"

As for Schrödinger's equation, it relates the advancement in time of the amplitude and phase of a wave function to their rate of change in space. Inserting an arbitrary set of misalignments in the phase (alternative 2) will certainly alter these rates of change and likely spoil the equation. How can Schrödinger's equation be satisfied if we insist on meddling with the phase this way? It cannot, *unless the equation contains other terms that cancel the effect of the misalignments.* There is no other place for such terms to come from (in alternative 2) except the gauge-transformed vector potential. Therefore, Schrödinger's equation *must* contain terms that depend on the vector potential, and those terms must cancel effects of the arbitrary misalignments in phase. The remarkable conclusion is that demanding the equations of motion be insensitive to arbitrary choices in the reference frame for phase *requires* the presence of the electromagnetic potential field in the equations for the motion of the electron wave function. Mathematically speaking, the appropriate terms to be added to Schrödinger's equation can be found easily. Maxwell's equations have to fit Schrödinger's equation like a glove to make all this work. [27]

The situation is analogous to choosing an arbitrary frame of reference for particle motion at each point of space-time in Newtonian physics. Einstein worried about choices that "jump around" while you measure, so you might think you are measuring a real force when you are really measuring the acceleration due to your own jumping. A real gravitational force, however, according to Einstein, will show up as an *intrinsic* curvature in space-time that produces an intrinsic curvature in the world-lines of matter, regardless of what coordinate system is used to describe them, even a silly system that jumps around. Gauss and Riemann gave us tools to calculate this curvature from measurements on a world-line in *any* coordinate system. But we are not talking about trajectories in space-time here, we are talking about the phase of the Schrödinger wave.

If the electromagnetic field did not exist, you could not adjust its potential with a gauge transformation to get rid of the artificialities introduced by arbitrary changes in the reference angle for quantum phase. Thus *the phenomenon of electromagnetism is needed to ensure that predicted events are independent of arbitrary choices of reference phase for the quantum wave function.* Weyl argued that if a quantum phase exists whose reference origin at each point is irrelevant, then there *must* be an electromagnetic field and it *must* obey Maxwell's equations.

What an extraordinary result! We now realize that Schrödinger's greatest contribution to science was the discovery of quantum phase as a new feature in the underlying geometry of nature.[28] Quantum phase is like a little dial attached to each point in space. If there is no electromagnetic field present, the zero marks on the dials can be arranged so all the pointers indicate the same angle. But if there is a field present, try as you might, you cannot get them all to read the same, just as you cannot trace a straight line on a globe. The field, or rather its vector potential, provides an intrinsic connection among the set of dial pointers at different points in space-time. Geometers call this kind of structure a *fiber bundle*, each dial representing a *fiber*, and the entire set of them over all the space-time points comprising the *bundle*. What physicists call the *vector potential*, mathematicians call the *connection* over the bundle. The geometry of our world, in the vision of Weyl–Schrödinger (thus far in our story), is therefore that of a one-dimensional fiber bundle over the four-dimensional manifold of space-time. Electromagnetic forces and gravity are "explained" as distortions in this geometry associated with the distribution of energy, mass, and electricity throughout the universe.

Unfinished business. There are problems, however, with this simple picture. One problem is that no one has yet been able to incorporate the force of gravity consistently into the quantum framework. In Einstein's theory the components of the metric play the role of potentials for the gravitational force and can carry waves, just like electromagnetic waves, that possess energy and momentum (but are expected to be very weak in our tame corner of space-time). The same arguments about thermal equilibrium that forced Planck to introduce a relation between the energy and frequency of light waves apply to gravity waves. If Einstein's picture is right, then there must be a quantum theory of gravity. Unfortunately, when the logical path that led to a successful quantum theory of electromagnetism is followed for Einstein's gravity, it leads to serious technical problems that I will not attempt to explain here. Some important results have been obtained, and intriguing lines of investigation are being explored, but a credible quantum theory of gravity does not yet exist.

Gravity is completely negligible at the scale of atoms and nuclei. So we should not need to incorporate it in a theory devoted to microscopic matter.[29] But the simple quantum theory fails for other more immediate reasons. It does not take into account the remarkable complexity of the microscopic world. In particular, it does not account for the forces that cause the binding and decay of the nuclei of atoms. After the spectacular successes of the 1920s, another half-century of meticulous observation and analysis passed before a coherent description of microscopic structure began to emerge. That description, the Standard Model, builds on the foundations explained in this chapter plus many additional diverse pieces. All are tied together with mathematical glue, the essential ingredients of which must be our next topic. Fortunately, the really important topics are easily grasped with little technical background.

NOTES

1. *Mendeleev's periodic table.* The table of chemical elements can be found in many convenient forms on the World Wide Web (search for *periodic table*). A similar, much larger table exists for the *nuclei* of elemental atoms (search for *table of nuclides*). Two tables are needed because atoms formed on different nuclei have nearly the same chemical properties if the nuclei have the same electric charge. See Chapter 7.

2. *Energy in the electromagnetic field.* If the magnitude of the electric force per unit charge is E, and the corresponding magnetic quantity is B (Maxwell reserved "M" for other purposes), then the total electromagnetic energy in a small volume ΔV is proportional to $(E^2 + B^2)\Delta V$. The coefficient of proportionality depends on the size of the units used to measure the charges. The fields E and B are understood to be evaluated within the small volume,

and there is no limit on their size. *Energy in light waves:* A light wave has both electric and magnetic fields which always point at right angles to each other and to the direction of the wave. In the simplest case (so-called linear polarization) the fields are in phase. A stationary observer would see the energy peak twice in each wavelength, once at the crest and again at the trough. Light intensity meters smooth over these peaks to register the average energy per cycle.

3. *Measurements of black body radiation.* Careful measurements of the black body spectrum were made in Germany at the end of the nineteenth century, particularly by O. Lummer and E. Pringsheim in Berlin. The spectrum of background cosmic microwave radiation that bathes Earth from every direction – discovered by A. A. Penzias and R. W. Wilson in 1964 (Nobel Prize 1978) – follows precisely the same black body curve with a temperature of 2.725 K. See Weinberg (1993) (Note 32 in chapter 2).

4. *Planck's formula.* Planck reported his formula on December 14, 1900 at a meeting of the German Physical Society. The formula gives the energy $u(f)$ per unit volume V per unit interval of frequency near f, so that $u(f)\Delta V\Delta f$ is the actual energy of the electromagnetic field in a small volume ΔV in the small frequency range f to $f + \Delta f$. I quote Planck's formula here in full so you can see it is a definite and not even very complicated mathematical expression:

$$u(f) = (8\pi h f^3/c^3)/[\exp(hf/kT) - 1],$$

where c is the speed of light (meters per second) and T is the absolute temperature measured in units whose size is adjusted by k, called *Boltzmann's constant.* For Celsius or Kelvin units, k is the very small number $k = 1.38 \times 10^{-23}$ J per degree. Recall that temperature measures the average kinetic energy of thermal atoms. Boltzmann's constant converts conventional temperature units to conventional energy units. If temperature were measured directly in Joules, Boltzmann's constant would equal "one." The graph of this formula is the bell-shaped curve of Figure 3.1 with the peak at $hf = 2.82kT$. Thus you can infer the temperature by observing the peak (the brightest color) in the spectrum. The *exponential function* "exp" is defined in Note 2 of Chapter 4.

5. *Definition of energy.* The pre-quantum definition of energy is tied to Newton's law $F = ma$, which is now understood to apply to an approximate picture of Nature, valid when effects involving Planck's constant are negligible. The old theory places no restriction on the energy of a mechanical oscillator. In the new theory, a tuning fork whose tines oscillate accurately with frequency

f can possess only a discrete set of equally spaced energies (i.e., not every energy). The difference between one allowed energy and the next is the *quantum of energy hf*. Because h is so small, this stepwise behavior is completely obscured in the motion of large objects, which is one reason the discovery of quantum theory and the correct definition of energy was so long delayed.

6. *Photoelectric effect*. Einstein's 1905 picture of the photoelectric effect regards electrons as little particles that are bound to the surface of metals with a characteristic energy E_o. If light of frequency f consists of bits (*photons*), each with energy hf, then a bit can eject an electron if hf exceeds E_o, leaving the electron with energy $E = hf - E_o$. Thus the ejected energy does not depend at all on the intensity of the light, proportional to the square of the electric force in the light wave, but rather on its frequency. This relationship was finally confirmed by R. A. Millikan in 1916. Einstein received the Nobel Prize in 1921 for this work, Millikan in 1923 (in part) for its confirmation. Today we understand that neither photons nor electrons are "little particles." The subsequent mouse-trapping detection of the released *photo-electron* is observed as a macroscopic electric charge or current at the position of the detector.

7. *Bohr's theory of atomic spectra*. Bohr's model for the motion of the electron in an atom depended on then-recent discoveries by Ernest Rutherford. In 1911 Rutherford had established that nearly all the atomic mass is localized in a nucleus about 100,000 times smaller than the atom itself. He did this by directing alpha particles (helium nuclei) ejected from a radium source toward a thin metallic foil and detecting the fraction that were scattered at various angles. Some were scattered back toward the source, signifying a concentration of mass within the target. The size of the concentration could be determined from the angular distribution. Bohr studied with Rutherford in Manchester for a few months in 1912, and produced his analysis of the hydrogen spectrum about a year later. See Pais (1991) for a detailed description of the Bohr model.

8. *De Broglie's paper on quantum waves*. De Broglie was troubled by the fact that the Planck frequency $f = mc^2/h$ for a particle at rest was not well defined when viewed from a moving frame. Time dilation would cause the apparent frequency to *decrease*, but relativistic effects cause the inertial mass to *increase* with the speed of motion, and therefore the observed Planck frequency should increase. Which is correct? De Broglie's answer was to introduce a new phenomenon that had not only a temporal periodicity, but also a spatial periodicity – that is, a sinusoidal wave associated with the moving system. He

chose the wavelength to make the new frequency equal to the correct Planck frequency after correctly transforming both distance and time to the moving frame, à la Einstein. The important and fascinating point is that the wave nature of matter can be inferred by making Planck's relation relativistically invariant. Einstein could have seen this many years earlier, but didn't.

9. *Observations of de Broglie waves.* De Broglie presented his doctoral thesis in November 1924, to which Einstein drew attention in a paper that appeared in February 1925. In July, W. Elsasser published a note that used de Broglie's ideas to explain certain features of electron scattering that had been observed by C. Davisson and C. H. Kunsman in 1921. Davisson and L. H. Germer published a paper in 1927 clearly showing a diffraction pattern from electrons scattering on a nickel crystal. Davisson was working at the American Telephone and Telegraph Company laboratories, later to become the famous Bell Laboratories. Elsasser was a student of Max Born's at Göttingen. Other experiments were performed shortly after in Aberdeen and Tokyo. This episode illustrates how much communication there was among physicists, how closely experiment and theory were linked, and how international the effort was to understand microscopic phenomena during this period. The history is sketched in Jammer (1966).

10. *Schrödinger's wave equation.* In its nonrelativistic form, Schrödinger's equation expresses the rate of change in time of the electron field, or *electron wave function*, at a given point in terms of the magnitude and rate of spatial variation of the field near that point. This expression for the rate of change can be related in simple cases to the "classical" Newton's Law formula for the energy of the system written in terms of the positions and momenta of its particles. Schrödinger's theory links to classical mechanics by identifying the *rays* – curves that pierce the moving wave fronts perpendicularly – with classical particle trajectories. Schrödinger devoted much of his 1933 Nobel acceptance speech (available at www.nobel.se) to the relation between optics and particle motion. See also Note 11 in Chapter 4 and the Appendix for more on Schrödinger's equation.

11. *Wave-like aspects of mechanics.* Hamilton's goal was to provide a mathematical basis for optics that was as powerful and elegant as that for mechanics. During the eighteenth century, a series of great French mathematical physicists, notably Lagrange (1736–1813), had recast Newton's Laws in forms of successively greater abstraction. Hamilton developed a formal theory of optics that linked light waves with

light rays (see preceding note), and then proceeded to recast the formal theory of mechanics in similar language, but without referring to waves. De Broglie and Schrödinger showed that this optics-like structure of mechanics has its origin in the quantum field that obeys Schrödinger's equation.

12. *Heisenberg's matrix mechanics*. Heisenberg (1925) found a way to summarize the accumulating *ad hoc* recipes of quantum mechanics in a mathematical formalism that replaced symbols for quantities like position, momentum, and energy with *matrices* (see Chapter 4). Although visiting Copenhagen at the time of his discovery, he was a student of Max Born's at Göttingen, and subsequently Born, Heisenberg, and Pascual Jordan developed the full theory. They converted the equations in Hamilton's formulation of mechanics to matrix form and showed how quantum properties emerged. Pais (1991) and Jammer (1966) are convenient historical references for this material.

13. *Bohr invited Schrödinger . . .* This famous incident is captured as a vignette in Michael Frayn's play *Copenhagen* (Frayn, 1998). This part of the play is based on memoirs of the participants:

Heisenberg (to Bohr)	You invited Schrödinger here. . .
Bohr	To have a calm debate about our differences.
H.	And you fell on him like a madman. You meet him at the station – of course – and you pitch into him before he's even got his bags off the train. Then you go on at him from first thing in the morning until last thing at night.
B.	*I* go on? *He* goes on!
H.	Because you won't make the least concession!
B.	Nor will he!
H.	You made him ill! He had to retire to bed to get away from you!
B.	He had a slight feverish cold.
H.	Margrethe had to nurse him!
Margrethe (Bohr's wife)	I dosed him with tea and cake to keep his strength up.
H.	Yes, while you pursued him even into the sickroom! Sat on his bed and hammered away at him!

14. *Born's statistical interpretation.* Philosophically speaking, the statistical interpretation of Schrödinger's wave function is one of the most important aspects of quantum theory, and yet Born's contribution (for which he received a long-deserved Nobel Prize in 1954) is not widely appreciated. In a postscript to Frayn (1998) the author admits that "I have grossly understated the crucial role played by Born himself and by his pupil Pascual Jordan at Göttingen in formulating quantum mechanics. . ." The statistical interpretation can be traced back to a 1917 paper by Einstein, and had been clearly stated in the context of matrix mechanics by Born and Jordan in 1925 before Born restated it for the wave function after Schrödinger's papers appeared the following year.

    Problems arise when applying the statistical interpretation to the wave function that describes the entire universe, which is unavoidable in cosmology. Murray Gell-Mann includes a popular discussion of a "Modern Approach" that seeks to address these problems (Gell-Mann, 1994). See particularly the chapter *A Contemporary View of Quantum Mechanics.* The approach described by Gell-Mann, known as the *consistent histories* view, is an extension of the interpretation presented here and leans heavily on the concept of decoherence described below in Note 17. A more complete but technical account of this approach, written at the undergraduate level, is Omnès (1994).

15. *Reality a social phenomenon.* These paragraphs are the tip of a very large philosophical iceberg. Most physicists have little patience with philosophical discussions of what is real, which usually entail technical issues of linguistics and psychology. Steven Weinberg's essay *Against Philosophy* (Weinberg, 1992) is an articulate expression of this position. In view of the special character of quantum theory, it is necessary at some point to confront the issue of what we mean by "real." In this book I use the word as I think Bohr meant it, namely to apply in its ordinary usage only to macroscopic phenomena. The mainstream physicist who most constructively encouraged deep thinking about these issues in the Bohr tradition is John Archibald Wheeler (1911–2008), whose autobiography is accessible to the non-specialist (Wheeler, 1998). Somewhat more demanding, but especially lucid and penetrating, are the collected philosophical papers of John S. Bell (1928–90) (Bell, 2004). See also the philosophical sections of Omnès (1994). If you want to know where real philosophers stand, I recommend looking into works by the winners of the

Lakatos Award in the Philosophy of Science, sponsored by the London School of Economics and listed on their website. For a brief and relatively painless starter, try Redhead (1995).

16. *Schrödinger's cat.* "Reality resists imitation through a model. So one lets go of naïve realism and leans directly on the indubitable proposition that *actually* (for the physicist) after all is said and done there is only observation, measurement. Then all our physical thinking thenceforth has as sole basis and as sole object the results of measurements which can in principle be carried out, for we must now explicitly *not* relate our thinking any longer to any other kind of reality or to a model" (Schrödinger, 1935). A common misinterpretation of Schrödinger's cat example suggests that the wave function *does* represent reality and the cat does somehow exist in limbo until the box is opened and the state of the cat observed by an intelligent observer. See the following note.

17. *The line between "detector" and "environment."* John von Neumann demonstrated the need for such a boundary in his classic study of the quantum measurement process (von Neumann, 1955). Some have argued that the only logical place to draw the line is at the point of cognition by an intelligent observer. I think this only works if *I* am the observer, because everyone else is part of my environment. This solipsistic conclusion gives up on reality as a socially shared experience. Bohr insisted that we draw the line much closer to microscopic phenomena, which is the view I adopt here as closest to the Copenhagen spirit. As described in the text, the phenomenon of decoherence establishes a natural minimum for $t_2 - t_1$.

18. *Decoherence.* "The mechanisms of decoherence are different from (though related to) those responsible for the approach of thermal equilibrium. In fact, decoherence precedes dissipation in being effective on a much faster timescale, while it requires initial conditions which are essentially the same as those responsible for the thermodynamic arrow of time" (E. Joos, in Giuliani *et al.*, 2003, Introduction). This is a technical monograph, but parts should be comprehensible to nonspecialists. A more accessible reference is Omnès (1994). The best popular reference I have found is Lindley (1996). Christopher Lehmann-Haupt, in his review of this book for the *New York Times*, says "matters become less clear when he [Lindley] defines the instrumental concept of 'decoherence.'" I agree, but the discussion here as elsewhere in Lindley's book matches the theory with admirable integrity.

19. *Quantum particles.* In his popular 1983 lectures at the University of California at Los Angeles on quantum theory, Richard Feynman said in the first lecture that "I want to emphasize that light comes in this form – particles. It is very important to know that light behaves like particles, especially for those of you who have gone to school, where you were probably told something about light behaving like waves. I'm telling you the way it *does* behave – like particles." But in the third lecture, he said "It's rather interesting to note that electrons looked like particles at first, and their wavish character was later discovered. On the other hand, apart from Newton making a mistake and thinking that light was 'corpuscular,' light looked like waves at first, and its characteristics as a particle were discovered later. In fact both objects behave somewhat like waves, and somewhat like particles. In order to save ourselves from inventing new words such as "wavicles," we have chosen to call these objects "particles," but we all know that they obey these [quantum] rules ... It appears that *all* the "particles" in Nature – quarks, gluons, neutrinos, and so forth – behave in this quantum mechanical way" (Feynman, 1985). I discuss Feynman's ideas briefly below, in Chapter 7. See also my article *What is a Photon?* (Marburger, 1996).

20. *Superposition of waves.* Joseph Fourier (1768–1830) discovered early in the nineteenth century that a wave of any shape can be represented by a sum of pure sine-waves, each with a unique amplitude, phase, and wavelength. This was important because the sine-wave representation gives a good approximation for the *entire shape*. Analysts prior to Fourier, particularly Isaac Newton, represented complicated shapes with so-called power series expansions which approximate arbitrary shapes accurately only *near a single point*. Other families of shapes can also be superposed to represent arbitrary waves, and many of them are useful in quantum mechanics.

21. *Uncertainty relation.* Heisenberg did not actually derive the rigorous inequality from the mathematics of quantum theory in his famous 1927 paper, but asserted that it could be done, and included a rigorous treatment in lectures at the University of Chicago about a year later. Weyl also derived it in his book on group theory in quantum mechanics which came out at about the same time. The earliest attempt at a derivation, cited by Heisenberg in his lectures, was published by E. H. Kennard shortly after Heisenberg's proposal. It is an unusual proof, presented somewhat cryptically and ignored by subsequent authors, including Heisenberg

(Marburger, 2008). Classic sources on the role of the uncertainty relations in the interpretation of quantum mechanics are Niels Bohr's chapter *Discussion with Einstein on Epistemological Problems in Atomic Physics* in Schilpp (1959) (this book also includes Einstein's lucid discussion of the statistical interpretation), and Heisenberg (1930). From a modern perspective, these accounts are more useful for demonstrating the consistency of quantum mechanics than in illuminating the deep issues of interpretation. The Bohr chapter is reprinted in Wheeler and Zurek (1983), which includes more recent material on these issues.

22. *Energy–time uncertainty relation.* Planck's relation identifies energy with frequency, and frequency, like wavelength, is unique only for a phenomenon that recurs "forever." The time duration of a phenomenon is related to its spectrum of frequencies in the same way the spatial extent of a phenomenon is related to its spectrum of wavelengths. This has implications for the interpretation of the "law" of conservation of energy: energy (frequency) is strictly conserved only for processes that run undisturbed forever. The physical interpretation of the resulting energy–time uncertainty relation is more subtle than the position–momentum relation because time and position variables play different roles in the (nonrelativistic) theory. Y. Aharanov and D. Bohm analyzed the situation in a 1961 paper reprinted in Wheeler and Zureck (1983) and concluded that despite the mathematical "uncertainty," "energy can be measured reproducibly in an arbitrarily short time."

23. *Incompatible with astronomical observations.* Zero-point energies of quantum fields should contribute to a term in Einstein's equations for the geometry of space called the *cosmological constant*. Einstein originally inserted this term to cancel the expansion of the universe before Hubble's observations showed expansion to be a real phenomenon. Current observations suggest the need for a cosmological constant, but nothing close to what would be predicted by quantum zero-point energies which is more than a hundred orders of magnitude too large! D. W. Sciama's paper in Saunders and Brown (1991) is a rich but somewhat technical source of information on zero-point energies. A good popular treatment is Barrow (2000).

24. *Quantum phase.* Although he never uses the word "phase," Richard Feynman explains the concept of quantum phase beautifully (Feynman, 1985). David Bohm, in his well-known introductory textbook, analyzes the

possibility of alternatives to Schrödinger's equation that do not require a complex wave function, with the conclusion "We are thus required, in the non-relativistic theory, to use an equation that is of first order with regard to the time, with complex wave function" (Bohm, 1951, section 4.5). For further information on phase and its representation by complex numbers, see the following Chapter 4 and particularly Note 2 in that chapter.

25. *Weyl's idea applied to quantum phase.* It is not possible to grasp the significance of the Standard Model without an appreciation of Weyl's idea. Nonmathematical treatments are rare and difficult to follow. Crease and Mann (1996) give a brief account, and translations of original papers may be found in the technical history (O'Raifeartaigh, 1997). Schrödinger had an early insight into the quantum significance of Weyl's theory, which caught the attention of Fritz London, who made the idea explicit (London, 1927). Weyl immediately embraced London's interpretation and elaborated the theory in further important papers.

26. *Reality of the four-vector potential.* A physical manifestation of the vector potential can be detected through its influence on the quantum phase. In 1959 Y. Aharonov and D. Bohm analyzed a hypothetical experiment in which a long magnetic coil (a *solenoid*) is placed in a stream of electrons. In this arrangement the magnetic field is confined to the interior of the coil, but the vector potential extends outside the coil, its direction curling around so on one side it points along the electron flow, and opposes it on the other side. The Schrödinger phase is thus retarded on one side relative to the other so there is constructive or destructive interference downstream from the coil, depending on the magnetic flux in the coil (proportional to the current in the solenoid) (Aharonov and Bohm, 1959). Many experiments since have demonstrated the effect. The point is that the electrons, being entirely outside the solenoid, are never exposed to the magnetic force, but are nevertheless affected by the vector potential. As T. T. Wu and C. N. Yang put it in their excellent review article, "[The electric and magnetic fields] together do not completely describe the electromagnetic field, and ... the vector potential cannot be totally eliminated in quantum mechanics" (Wu and Yang, 2006).

27. *Invariance of the electromagnetic force to gauge transformations.* To be precise, Maxwell's equations imply that the flux of the electromagnetic field through an arbitrary surface in four-dimensional space-time equals the length of the perimeter of the surface, as gauged by the four-vector

potential and the charge. In the absence of forces, the vector potential can be set to zero so the gauged length of the perimeter is zero. Schrödinger's equation must be such that the phase of the wave function has a contribution equal to the gauged perimeter.

28. *Significance of quantum phase.* According to C. N. Yang, Paul Dirac had this to say about the significance of phase: "So [said Dirac, at age 70] if one asks what is the main feature of quantum mechanics, I feel inclined now to say that it is not non-commutative algebra. [As employed in Heisenberg's matrix mechanics.] It is the existence of probability amplitudes which underlie all atomic processes. Now a probability amplitude is related to experiment, but only partially. The square of its modulus is something that we can observe. That is the probability which the experimental people get. But besides that there is a phase, a number of modulus unity which can modify without affecting the square of the modulus. And this phase is all important because it is the source of all interference phenomena, but its physical significance is obscure. So the real genius of Heisenberg and Schrödinger, you might say, was to discover the existence of probability amplitudes containing this phase quantity which is very well hidden in nature and it is because it was so well hidden that people hadn't thought of quantum mechanics much earlier." Quoted in Yang (2003).

29. *Gravity negligible for microscopic matter.* The uncertainty relation, however, associates large momentum fluctuations Δp with small scales Δq, and therefore at some tiny scale the energy density associated with these fluctuations must lead to general relativistic spatial curvatures on the order of Δq itself. At this point the theory of gravity must be joined with quantum theory.

4 The language of Nature

Schrödinger's equation and the "wave function" that satisfies it are the first bold brushstrokes of a new portrait of Nature that physicists gradually filled in during the middle half of the twentieth century. The emerging pattern was strange, but not entirely alien to science. It was drawn, to paraphrase Galileo, in a mathematical idiom "without some knowledge of which it is humanly impossible to understand." While nonmathematical accounts are conceivable, I would rather at this point say more about the mathematics because it will deepen our view of the whole enterprise of modern physics. Most people never get beyond the "math is numbers and numbers are boring" barrier that hides behind it a limitless universe of beautiful things. So let us delve briefly into the nonboring side of mathematics.

4.1 MATHEMATICAL THINGS

During the later nineteenth century, physicists awakened gradually to a curious fact: many natural phenomena are most easily described using objects from the abstract world in which mathematicians amuse themselves. This is the world that contains the rules of chess, "spaces" of N dimensions, and various extensions of the number system that go far beyond anything required for counting. In practical life most of us use mathematics to manipulate *quantities* (amounts). The letters in high school algebra stand for measures of things like distance D (number of miles), cost C (number of dollars), or temperature T (number of degrees Fahrenheit or Celsius). This use of symbols to stand for simple numbers is just the tip of the iceberg of mathematical objects. Sadly few of us have, or seize, the opportunity to learn about the wonderful world of other things in the mathematicians' universe. Nineteenth-century mathematicians invented new conceptual objects and new operations among them that have only a remote family resemblance to the numbers of our everyday experience.[1]

Arrays. Some of the new objects represent *arrays* of numbers. For example, a square tic-tac-toe array (a *matrix*, see below) of nine numbers might be treated as a single object, and symbolized in a formula by a single letter. "Addition" or "multiplication" of two such objects would be defined by a set of made-up rules applied to the numbers they contain. Arrays are familiar to anyone who has set up a computerized spreadsheet in which columns of numbers are treated like single entities: Add column A to column B to get column C. Then in the formula $A + B = C$, the letters do not stand for single numbers, but for entire columnar arrays of numbers. "Addition" in this case means "for each row, add the number in column A to the number in column B and place the sum in column C."

Complex numbers. The amplitude–phase pair (a, θ) of Schrödinger's electron field can be treated as a single object called a *complex number* and represented in the theory by a single symbol (often the Greek letter *psi*: ψ). So $\psi = (a, \theta)$. Mathematicians have invented an operation on complex numbers corresponding to multiplication: The amplitude of the "product" of two complex numbers is the ordinary product of the amplitudes (one times the other), and the phase of the "product" is the *sum* of the phases. If you write the amplitude–phase pair as (a, θ), then you could represent the product of two such pairs as $(a, \theta) \cdot (b, \phi) = (ab, \theta + \phi)$, where the dot, usually omitted, reminds us that this is a kind of multiplication. The "magnitude" of a complex number, written $|\psi|$, simply equals its amplitude a, a positive number.

A kind of "sum" can also be defined for complex numbers: Draw arrows pointing in the direction of the phase angles with lengths equal to the amplitudes. Then move the arrows – in any order but without changing their lengths or angles – so the tail of one touches the tip of another. A new arrow that stretches from the first tail to the last tip is defined to have the amplitude and phase angle of the "sum." Since multiplication can be viewed as repeated addition (e.g., $3 \times 5 = 5 + 5 + 5$) care has been taken to have the "sum" consistent with the "product."

Despite risk of confusion, the usual signs (+, –, ×, ·, . . .) are used for their new invented counterparts for complex numbers and for other new objects considered below. Note 2 gives more detail that you might want to consult later.

A complex number whose phase advances with time is represented by an arrow that winds around like a hand on a clock, which is useful for describing waves and oscillating systems as in Figure 3.5. Other properties of complex numbers make some kinds of mathematical calculations easier. So they are often introduced artificially in formulas for things properly described by ordinary numbers (electric fields, for example) as a first step in a purely mathematical manipulation of the formulas. Schrödinger's field was the first instance of an object in physical theory that is *intrinsically complex*.[2]

Vectors. By the end of the nineteenth century, physicists were using single symbols like **x** to refer to the arrays of three coordinates (*x*, *y*, *z*) that locate objects in space (such as distance up from the floor and out from two perpendicular walls), or to the three components of velocity or force along horizontal and vertical directions in a frame of reference. Newton's Laws and Maxwell's equations became simpler when expressed in the new concise notation. People began to think of all the components together as representing a single thing, a *position vector* for example, or a *force vector*. The word *vector* describes a geometrical object that can be represented by several numbers that act like the coordinates of length, width, and height of a point in space referred to a standard frame of coordinate axes. When there are two or three components, it can be represented by an arrow as in Figure 4.1,

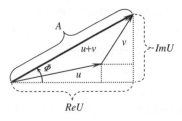

FIGURE 4.1 The operation corresponding to addition of two complex numbers *u*, *v* can be visualized with arrows representing the number pairs. Here $U = (A, \Phi) = u + v$. The dotted lines are *projections* of the arrows on horizontal and vertical axes. The projections are called "real" and "imaginary" parts of the corresponding complex numbers.

where the vertical and horizontal components are shown as dotted lines. The component representation of a vector is a special case of the arrays that form the columns of a spreadsheet. Complex numbers act like vectors in a flat plane (i.e., two dimensions) called the *complex plane*. There is no new content here, just a different way of speaking about things. But there is a message in the medium of expression. When Minkowski drew attention to the fact that the natural objects in Einstein's new physics were vectors in a space of *four* dimensions, one of which was time, the mathematical language scientists were using prepared them to see deep implications in the idea.

Notation. Even if you knew nothing of mathematics, you would notice immediately the difference between a nineteenth-century scientific work – Maxwell's famous *Treatise on Electricity and Magnetism*, for example – and a modern textbook. To cite a random example, where Maxwell wrote

> The quantities a, b, c are the components of magnetic induction, and are related to α, β, γ, the components of magnetic force, by the equations given in Art. 400,
>
> $$a = \alpha + 4\pi A,$$
> $$b = \beta + 4\pi B,$$
> $$c = \gamma + 4\pi C.$$

we would today use bold symbols for the vectors: $\mathbf{a} = (a, b, c)$, $\boldsymbol{\alpha} = (\alpha, \beta, \gamma)$, etc. and write

> Magnetic induction $\mathbf{a}$ is related to magnetic force $\boldsymbol{\alpha}$ by the equation of Art. 400,
>
> $$\mathbf{a} = \boldsymbol{\alpha} + 4\pi\mathbf{A}.$$

Not only is the modern version shorter, it is also easier to read. We cannot help but think the "real thing" is the vector $\mathbf{A}$, not the array (A, B, C) for which it stands. Besides, the vector notation is indifferent to the particular reference frame in which the components are measured, which Einstein says is irrelevant anyway. No wonder we say today that

"magnetic force *is* a vector" rather than "the components of the magnetic force *may be represented by* a vector." Both statements are true,
but the latter drags in the particular observer's viewpoint through the
components, which depend on the frame of reference. In the example
above, if the x-axis of the reference frame were along the magnetic
force, then α would equal the force magnitude, and β and γ would be
zero in the equations above, but choosing the y-axis along the force
would make α and γ zero, and β would equal the force magnitude.
Vector components are often indicated by subscripts: $\mathbf{A} = (A_x, A_y, A_z)$.

Matrices. You may have skipped over the previous paragraph
with a shudder, recalling the dreaded *simultaneous equations* from a
high school algebra course. Array notation compresses simultaneous
equations into single, simple-looking equations. Here is another fright-
ening example:

$$a = A\alpha + A'\beta + A''\gamma,$$
$$b = B\alpha + B'\beta + B''\gamma,$$
$$c = C\alpha + C'\beta + C''\gamma,$$

which relates the array we have been calling $\boldsymbol{\alpha} = (\alpha, \beta, \gamma)$ to the new array
$\mathbf{a} = (a, b, c)$. But the whole thing can be simplified if we define a new tic-
tac-toe (matrix) array $\mathbf{M}$:

$$\mathbf{M} \;=\; \begin{matrix} A & A' & A'' \\ B & B' & B'' \\ C & C' & C''. \end{matrix}$$

.

Then just *define* the symbol $\mathbf{M \cdot \boldsymbol{\alpha}}$ to stand for the right-hand side of
the messy-looking set of three equations so we can replace them all
with the simple-looking equation $\mathbf{a} = \mathbf{M \cdot \boldsymbol{\alpha}}$. Now you can think up
relations like $\mathbf{b} = \mathbf{N \cdot a} = \mathbf{N \cdot M \cdot \boldsymbol{\alpha}}$, which imply a sort of "multiplication" between the matrices $\mathbf{M}$ and $\mathbf{N}$. Just as in ordinary algebra,
the dot symbol for the new made-up multiplication operation is
often suppressed, so $\mathbf{NM}$ is understood to mean $\mathbf{N \cdot M}$. The rules of
matrix multiplication are not sent from heaven. They are invented
by mathematicians precisely to represent the complicated-looking
equations involving individual components of arrays. But Nature's

phenomena happen to be summarized in their simplest and most general form using symbols that obey these made-up rules.

What mathematicians love about arrays is that they suggest new concepts you would not think of when a symbol stands only for a single number. For example, each array has a *dimension* (number of rows or columns in a spreadsheet, or number of numbers in the array). You can also invent lots of new numbers associated with arrays, such as the sum of the squares of all the numbers in the array, or the average size of the numbers, the greatest or least of the numbers, the difference between greatest and least, and so forth. Some of these new numbers have interesting properties, some do not. Mathematicians judge each other by their ability to invent new concepts like this that *are* interesting.

The game of objects and operations. The game is to start with numbers (1, 2, 3, ...) and operations on them (+, −, ×, ÷, ...), and then replace the numbers with other abstract objects like arrays, and invent generalizations of the operations that obey more or less the same rules as addition, etc. on numbers, but that apply to the new objects. Or you could invent entirely new operations that are not at all like the familiar arithmetic ones. The objects do not have to be numbers, or even stand for numbers. They could be colors or shapes, family members or grocery items. And the operations could be anything you like that produces some change – a color to its complement, for example, or a shape to its mirror image, a son to a daughter, or an apple to an orange. The point is that there are two kinds of things in mathematics – rather like the categories of nouns and verbs – one the material on which the other acts. From such grammatical elements the statements of mathematics are constructed.

Even boring arithmetic like $2 + 5 = 7$ becomes (slightly) more interesting if you think of the symbol "2+" as an operation that shifts any number by two digits: $(2+)\cdot a = a + 2$. And you can think of the complex number pair $(1, \alpha)$ as an operator that shifts the phase of any arbitrary complex number by α using the complex "multiplication" rule: $(1, \alpha)\cdot(a, \phi) = (a, \phi + \alpha)$. Weyl's reasoning that electromagnetism is necessary in a quantum theory of matter begins with the idea that wave functions must be invariant under this operation. That is, the wave functions ψ and $(1, \alpha)\cdot\psi$ both lead to the same probability $|\psi|^2 = a^2$. We think of electromagnetism as a consequence of this invariance.

A more ingenious idea is to start with a collection of *operations* and make them the "objects," and then combine them using new operations that act on the original set of operations! It sounds circular, but an example makes it clear: The starting set of object–operations could be all the rigid motions you could apply to some symmetrical figure that would make it look the same after the motion is done. These are called *symmetry operations* of the object. Think of rotating a square through 90° about its center, or through 180° or 270°. Now you can think of "multiplying" two such rotations by doing them in succession. "Inverse" rotations would be in the opposite sense, and so forth (in this example an operation like addition is not defined). It is a very pretty game, and it leads to mathematical structures even more fascinating than the game of chess. More about this later.[3]

4.2 SCHRÖDINGER'S WAVE AS A SET OF VECTOR COMPONENTS

Born's statistical interpretation of the wave function $\psi(A)$ requires us, the observer, to specify a detector A (where A might stand for location and time) before we can evaluate the probability $|\psi(A)|^2$ that Nature will cause a registration in a detector with that specific tuning. This interpretation entails two disturbingly divergent intuitions: that there is an objective Nature "out there" independent of ourselves, but that what we observe of Nature inevitably depends upon our free and arbitrary choice of what to look for. They could be called the guiding principles of Einstein and of Bohr, respectively. Einstein's greatest discoveries followed from his search for a description of Nature's rules that makes no reference to an observer's arbitrary choices. The creators of quantum mechanics succeeded by incorporating observer choices into the very foundations of the theory. Remarkably, however, the conceptual structure of quantum mechanics does permit a formulation that neatly separates observer choices from an underlying independent "something." P. A. M. Dirac (1902–84) was among the first to call attention to this possibility, and he developed an efficient symbolic language that clarifies the issue. Dirac's quantum language treats the choice of what to observe in precisely the same way that arbitrary choices of coordinate frames are handled in the classical theories, but with an alarming new feature that I will describe. This language, for nearly 80 years the standard among physicists, adds nothing new to the substance of the theory, but establishes a profound conceptual framework that shapes our picture of microscopic nature.

Poets show us new ways of looking at old situations that cause us to rethink their meaning. To many physicists Dirac's *Principles of Quantum Mechanics*, first published in 1930, is a work of poetry.[4] It appeared after four years of intense creative effort by the founders of quantum mechanics, including Dirac himself, during which a clear theoretical picture gradually emerged. Dirac's book presents the result in a simple and compelling form. I vividly recall reading it straight through one long weekend at home before I began my graduate work, taking little time to eat or sleep, to my parents' consternation. In my undergraduate courses at Princeton I had already seen much of the same material, but Dirac's exposition made it clear how everything fit together. And it was all simpler than I had realized.

You can see that the symbol for the wave function $\psi(A)$ for a single detector located at A has two parts, ψ and A, that resemble the two parts of the symbol for the component of a vector along a specific coordinate axis, for example V_x (Figure 4.2). We could write it ψ_A to make the resemblance stronger. The "V" in the vector notation refers to something "objectively out there" like an electric force pushing a charge in some direction, and the "x" refers to, say, the direction of an edge on my desk. The symbol V_x denotes the component of force parallel to the edge. I need to know all three components to find the strength and direction of the vector $\mathbf{V}$ – a bold-faced symbol to remind me it is not a number, but "something really out there" independent of any particular choice of reference frame. Appearances are not

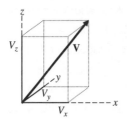

FIGURE 4.2A The components of the vector **V** are the lengths of the edges of a box (the coordinate "frame") whose longest diagonal is the vector. The shape of the box, and therefore the sizes of the components, depends on the arbitrary choice of directions of the coordinate axes x, y, z that define the edges of the box.

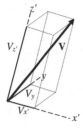

FIGURE 4.2B Rotating the coordinate frame – but not the vector – about the y-axis changes the x and z-coordinates, distorting the box but leaving the vector **V** unchanged. Here V_x changes to $V_{x'}$, and V_z changes to $V_{z'}$.

In Dirac's description of the wave function, the freedom to choose what to detect corresponds to the freedom to choose coordinate axes. The state of Nature under observation corresponds to the vector **V**. The wave function at position x is a single component of the state vector, as V_x is of **V**.

everything, but might it be possible to view the ψ in the wave function as a vector in its own "space," and the A as a label for a "coordinate axis" in this space corresponding to the arbitrary choice I made to set my detector there?

Hilbert space. This vector metaphor for the wave function runs deep. In the decades prior to the quantum discoveries, mathematicians had concluded that quantities like $\psi(A)$, i.e., mathematical *functions*, could indeed be regarded as components of a vector in a "space" *with an infinite number of dimensions*, one dimension, roughly speaking,[5] for each choice of A. If you know $\psi(A)$ for every A then you know everything you need to know about the underlying vector in the space. What makes the vector metaphor especially useful in quantum theory is a property of the quantum equation of motion (Schrödinger's equation) called *linearity*, which implies that the sum of any two solutions gives another valid solution. Two different wave functions $\psi(A)$ and $\phi(A)$, each of which describes a physically possible state of some system, can be combined just like vector components to make a third possible state: $\Psi(A) = a\psi(A) + b\phi(A)$. (The multipliers a and b have to be chosen to make the sum act like a probability amplitude, i.e., never exceeding "one".) We need this property to superpose waves as in Figure 3.6. You can therefore regard the symbols Ψ, ψ, and ϕ as referring to vectors in an *abstract vector space*. Call the entity whose components are ψ_A a *state vector* and for now designate it with a bold typeface Ψ, ψ, or ϕ in analogy with the notation for vectors in ordinary space. Dirac invented a more suggestive notation that I will describe below. When the great mathematician David Hilbert (1862–1943) realized how important it was for quantum mechanics, he developed the mathematical theory of this kind of space in detail, and we know it today as *Hilbert space* or *quantum state space*. Thus *the wave function at a point in space is a component of a vector in a new kind of "space" with an infinity of dimensions* in which different possible coordinate axes are labeled with different detector tunings.

This interpretation sheds light on the possibility of representing a wave function in different ways. As pictured in Figure 3.6, for example, you could represent $\psi(x)$ as a sum of pure oscillatory waves with different wavelengths L. You need to adjust the amplitudes and phases of each component wave so the sum comes out to the right $\psi(x)$ for each location x. Use the symbol $\psi^w(L)$ for the amplitude and phase of that particular component sine-wave with wavelength L (superscript w as a reminder for "wave"). Then we have *two* (infinite) sets of complex numbers that describe the same underlying vector: all the $\psi(x)$'s for the different positions x, and all the $\psi^w(L)$'s for the different L's. Regarded as graphs or functions of their arguments x and L, $\psi(x)$ and $\psi^w(L)$ are quite different. What is remarkable is that the two sets are related in the same way as the components of a vector expressed in two different coordinate systems. Each term in the sum of sine-waves adding to $\psi(x)$ is proportional to a $\psi^w(L)$ just as if the latter were a component of the *same* abstract vector ψ underlying $\psi(x)$. That is, if it makes sense to replace the notation $\psi(x)$ with the vector component symbol ψ_x, then it also makes sense to replace $\psi^w(L)$ with the symbol ψ_L *omitting the distinguishing superscript w* (because ψ is the same in both cases). The key idea here is the passage from a point of view that features *functions* – which are equivalent to graphs of one variable, ψ, plotted against another, x (as in Figure 3.4 or Figure 4.4 below) – to a different point of view that emphasizes *vectors* where a new kind of mathematical object ψ has components ψ_x along coordinate axes labeled x, and ψ_L along a different set of axes labeled L in a multi-dimensional space (as in Figure 4.2 above). *The choice of what physical quantity to detect dictates the choice of coordinate frames (bunches of mutually perpendicular axes) in the quantum mechanical Hilbert space. The choice of which value of the quantity to detect fixes the particular coordinate axis within the frame.*

4.3 THE QUANTUM STATE VECTOR IS NOT OF THIS WORLD

Just as Einstein labored to free the laws of motion from the contingency of observer choice, Dirac's view of quantum theory frees the description of the microscopic state from the ultimate observer's choice of what to measure. That's the good news. The bad news is that ψ, the observer-independent entity in the theory, *does not exist in the world of space and time.* It is a vector in an abstract space the magnitude of whose component (squared) along a particular direction gives the probability of detecting an event with a detector corresponding to that direction (assuming such a detector exists). Thus there are *two* important directions for each observation, one for the *state of Nature*, the other for what we choose to observe, which you could call the *state of the detector*. If you assign a length of "one" to each, then the component of one vector that falls like a noon shadow upon the other depends only on the angle between the two (see Figure 4.3). The squared magnitude of this component is the probability that the chosen detector will click when Nature is described by the state vector. Physicists learn early in their studies how to calculate with vectors in two and three dimensions. The leap to vectors in infinite-dimensional Hilbert space may boggle the imagination, but the mathematical techniques remain familiar.

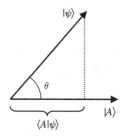

Dirac's notation. Dirac invented suggestive symbols for the state vectors and their components in Hilbert space that help to visualize the mathematical operations and ease the tedium of calculations. If Nature's state vector – what I called **ψ** – is written $|\psi\rangle$, and the vector along the dimension corresponding to the detector at A is called $|A\rangle$, then the projection of the first upon the second has the symbol $\langle A|\psi\rangle$. This is just the wave function, so $\psi(A) = \langle A|\psi\rangle$. Here A stands for a spreadsheet-like column of all the numbers or symbols required to specify or "tune" the detector – of which we will encounter many more later – and not just its location in space and time.

The world-state vector. Compare this description with our picture of macroscopic Nature as a dance of atoms whose world-lines weave a tapestry in time. Quantum mechanics imagine snapshots of microscopic Nature's state vector strung together to make a movie. The state vector will turn this way and that, winding its direction, not in three-dimensional space but in *Hilbert space*, rather like the arm of an (infinite-dimensional!) child pointing out the path of a butterfly. Its motion is determined by Schrödinger's equation.[6] This vector is a purely mental picture, however, an aspect of the model we have constructed that tells us something about the external world. It does not precisely mimic the external world itself. All we can actually observe of that world are clicks in a detector. Choosing the detector tells which direction to look in state space. If at some instant Nature's pointer coincides with that direction, you will *certainly* hear a click. If not, you *may* hear a click, and the probability of doing so equals (the squared magnitude of) the projection of Nature's pointer onto the detector's. If

Nature's arrow is perpendicular to that direction, then you will certainly hear no click at all. In the language of trigonometry, the probability for a click is the square of the cosine of the angle between the two directions, suitably generalized to make sense in a complex infinite-dimensional space (Figure 4.3).

The work of the quantum mechanic now divides into three parts. First, to discover the Rules of the Dance – Schrödinger's equation, or its successors that I will describe later. Second, to apply the Rules in specific cases, such as the emission spectrum of the hydrogen atom, or the radiative decay of a nucleus, where we know the conditions that fix a unique outcome. How quantum mechanics actually use the theory can be subtle, and I give some examples in the Appendix. The third task, to which much of quantum training is devoted, is to derive the properties of state vectors corresponding to specific types and arrangements of detectors (e.g., for position, momentum, angular momentum, energy, etc.). The work of prediction consists of finding the angle between the vectors for the state and for the detector. Usually you can answer useful questions without solving for the vectors first, taking a shortcut to calculate the desired component, or some other relevant quantity, directly. Professional physicists need to develop this problem-solving skill, but you don't.

4.4 A NEW PERSPECTIVE ON UNCERTAINTY AND COMPLEMENTARITY

Many aspects of quantum mechanics are "explained" by Dirac's vectorial picture. "Quantum uncertainty," for example, expresses the fact that Nature's state vector does not have to point along the direction of any particular detector vector. It will most likely have projections on many different detector directions, exactly as the vector V in Figure 4.2 has projections along three different coordinate axes. Corresponding to each projection is a probability of clicks – a number between zero and one equal to the squared magnitude of the projection. The spread of values over different directions (different detector tunings) is a measure of the uncertainty in the measured property.

If the state vector happens to point along a direction corresponding to a detector tuned to a definite *momentum*, it cannot simultaneously point along a *single* direction for definite *position* because no position direction aligns with any momentum direction in the space (see the following paragraph). Heisenberg's uncertainty relation, in this view, says less about the state of Nature than about how different coordinate frames are related in state space. Position and momentum are attributes that we choose to measure, not necessarily attributes always possessed in an objective sense by Nature.

The physical content of the uncertainty relation for position and momentum enters through de Broglie's formula. Write $|p\rangle$ for the vector corresponding to a detector tuned for momentum p and $|x\rangle$ for a detector tuned for position x. When Nature's vector points along $|p\rangle$ its wave function must be proportional to $\psi(x) = \langle x|p\rangle$. It makes no difference what physical system might have this momentum, but you can think of it as an atom moving freely so the momentum is constant. According to de Broglie the atom must be associated with a wave in space whose amplitude is the same everywhere and whose wavelength L is h/p. The phase of such a wave is simply x/L times $360°$ (or 2π radians), and its real and imaginary parts (Note 2) look like cosine and sine-waves. This special wave determines the relation between the position and momentum coordinate frames in the quantum Hilbert space because $\langle x|p\rangle$ is simply related to the "angle" between $|x\rangle$ and $|p\rangle$ (suitably generalized for Hilbert space). Because the magnitude of $\langle x|p\rangle$ never vanishes, $|p\rangle$ has a component along every $|x\rangle$ and *vice versa*. This is all the information you need to derive Heisenberg's relation (which requires some calculus).

By contrast, the wave function for a state of Nature with well-defined position a, can be written $\langle x|a\rangle$, which must vanish for all positions not close to a. Its graph versus x looks like a spike where $x = a$.

Detectable physical properties like position and momentum that correspond to different frames of reference in Hilbert space are called *complementary*.[7] (The frames must be truly different, not just differing in how their axes are labeled.) A state of Nature $|\psi\rangle$ expressed in one such frame possesses a wave function that is said to *represent* $|\psi\rangle$ in

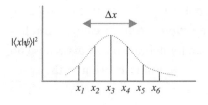

FIGURE 4.4A A plot of the squared magnitude of the projections of the state vector $|\psi\rangle$ on a sequence of directions in Hilbert space corresponding to different detector positions $|x_1\rangle$, $|x_2\rangle$, $|x_3\rangle$, ... This is a graph of probabilities for finding the system at x_1, x_2, x_3 ... and its spread is measured by Δx. The function $\langle x|\psi\rangle$ is called the *position representation* of the state vector.

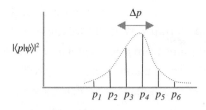

FIGURE 4.4B A similar plot for the projections of $|\psi\rangle$ on directions corresponding to detectors tuned to measure momenta $|p_1\rangle$, $|p_2\rangle$, $|p_3\rangle$, etc. produces a different function $\langle p|\psi\rangle$ called the *momentum representation* of the same state vector. The spread of the squared magnitude of this function is measured by Δp.

The relation between the directions $|x_1\rangle$, $|x_2\rangle$, $|x_3\rangle$ ... and the directions $|p_1\rangle$, $|p_2\rangle$, $|p_3\rangle$ ... in Hilbert space imposes a condition on the product of Δx and Δp for any $|\psi\rangle$, namely the *Heisenberg uncertainty relation* $\Delta x \, \Delta p \geq h/4\pi$.

that frame (Figure 4.4). Two different representations of the same state of Nature are related to each other through rotations of the frames of reference as in Figure 4.2.

What Einstein hated about this view of Nature was, first, the inherently probabilistic view of Born's statistical interpretation ("God playing dice"), but, second, that it gives too much power to the observer. Einstein thought things like position, momentum, and spin should be properties of physical reality independent of the observer, not just approximately for macroscopic systems, but exactly at every level. Today we view these properties as realized *jointly* in the observing apparatus and the microscopic stuff – two arrows, one for Nature, the other for my choice of what to measure, thrust out into an infinite-

FIGURE 4.5A Regarded as a function of x, the projection of the "momentum state" $|p\rangle$ upon $|x\rangle$ is a complex number whose real and imaginary parts are sinusoidal waves.

FIGURE 4.5B The projection of the "position state" $|a\rangle$ upon $|x\rangle$ vanishes except near $x = a$.

dimensional space. The result of my interaction with the universe, as far as I can guess it in advance, depends only upon the angle between the two. For Einstein, this was one arrow too many.

The idea of representing Nature as a vector in Hilbert space, however, removes much of the mystery about the arbitrariness of the time of detection in the discussion of Schrödinger's cat example in the previous chapter. During the initial "decoherence" period of evolution, it is difficult to imagine how one could build a detector (a macroscopic apparatus) that could capture the complexity of the quantum state vector. After decoherence caused by interaction of a microscopic system with many "lumpy" degrees of freedom in the environment, the state vector has separated into a set of distinct possibilities, each with its own probability of occurrence, that we can probe at leisure with a macroscopic detector of our devising. Most physicists would prefer to think of reality as represented by an infinite-dimensional vector than by a simpler structure in an infinite number of parallel universes as envisioned by Everett.

4.5 MORE STRUCTURE FOR SCHRÖDINGER'S WAVE: "INTRINSIC SPIN"

Neither of the notations $\psi(x)$ or $\langle x|\psi\rangle$ for Schrödinger's electron wave function gives the all-important phase its due. Quantum phase is really an additional structure – ultimately a *geometrical structure* – attached

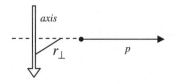

FIGURE 4.6 Angular momentum is a "kind of" vector defined with respect to an arbitrarily chosen perpendicular axis. Its direction is by convention the direction of advance of a right-hand screw turned by the momentum around the axis.

by the complex wave function to each point of space. Through Weyl's gauge argument the circular geometry of this structure determines the behavior of the electromagnetic field and how it links to Schrödinger's equation. One aim of this book is to show how Nature demands more such structures with more complicated "internal" geometries than the circle of quantum phase. The next simplest of these additional features is *spin*, whose introduction in the early history of our subject paved the way for other more profound internal structures – although spin itself is more subtle than it seems.

Angular momentum. Bohr's atomic model showed that rotational motion, or more accurately *angular momentum*, is quantized. Angular momentum is a measurable quantity in Newtonian physics equal to ordinary momentum p times the distance $r_\perp$ to an arbitrary axis perpendicular to p (Figure 4.6). Specifying it requires the magnitude $pr_\perp$ and the direction of the axis, so angular momentum is a kind of vector ("kind of" because it includes an arbitrary convention that links direction along the axis with the sense of rotation). When an electron orbits a nucleus in a circle the de Broglie wave wraps around an integral number of wavelengths $L = h/p$ as in Figure 3.10. That is, the circumference of an allowed orbit is $2\pi r_\perp = nL$, so the angular momentum $pr_\perp$ is $p(n/2\pi)L = p$ $(n/2\pi)(h/p) = nh/2\pi$ where n is an integer $(0, 1, 2, 3, \ldots)$. That is, if we could detect the angular momentum of electrons whirling within different atoms, we can expect to find values ranging from no angular momentum at all $(n = 0)$ to arbitrarily high integer multiples of $h/2\pi$.

An angular momentum detector requires settings for magnitude *and* direction. Magnitudes can be labeled by n as above. Given n, the detector will trigger on a set – a *spectrum* – of different values depending on which direction you choose for the detector axis. But here is a new

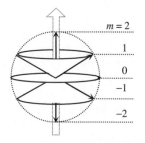

$m = 2$
1
0
-1
-2

FIGURE 4.7 The five "allowed" states for a total angular momentum of $2 \times h/2\pi$ ($n = 2$). The direction is only determined up to location on a cone, as shown. That is, only the *projections* of angular momentum on the detector axis are fixed for a given n.

phenomenon: *the directional spectrum is also quantized.* Specifically, for given n, only the values $mh/2\pi$ are ever observed for the component of angular momentum along any chosen axis, where m takes the values $(-n, -n+1, \ldots n-1, n)$, a total of $2n+1$ values in all. Imagine an angular momentum vector n units long tilted at a set of $2n+1$ different angles from the detector direction (Figure 4.7). For the values $n = (0, 1, 2, \ldots)$ the number of observable tilts is an odd integer $(1, 3, 5 \ldots)$. This result is a logical consequence (not an obvious one) of the fact that position and momentum are complementary variables (but see Chapter 6 for a deeper perspective). In the language of Dirac notation, the state corresponding to an angular momentum detector may be written $|n,m\rangle$, n for the magnitude, m for the direction of the detected angular momentum. The detector will only register an event if the electron state vector $|\psi\rangle$ has a component with total angular momentum $nh/2\pi$ and projection $mh/2\pi$. Then the probability of a registration at this tuning is $|\langle n,m|\psi\rangle|^2$.

Electron "spin". When examined closely, the spectrum of bright lines in the light emitted by excited atoms contains *multiplets* of closely spaced lines. Roughly speaking, *the number of lines in the multiplets are twice what the Bohr model predicts.* It turns out to be possible to relate this empirical "two-ness" in the spectrum to the notion that electrons themselves might spin like little tops. This "spin" adds to the angular momentum of the swirling electrons, but it is oddly different from the circulating motion analyzed above.[8]

Because it is electrically charged, a spinning or circulating electron would presumably act like a little electromagnet whose energy, and therefore the spectrum of light it emits when moving in an atom,

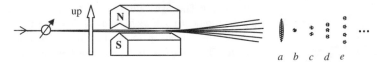

FIGURE 4.8 Spectra of possible spin directions for beams of various spinning objects: (a) expected classical spectrum; (b) no spin; (c) observed spectrum for electrons (least nonzero spin); (d), (e) ... observed spectra for higher spins. The incident objects are presumed to be randomly oriented. The vertical arrow determines "spin up," and the dots line up along this direction.

would be affected by a magnetic field. (This picture suggests the electron is like a little ball, which it is not, but this was the historical idea that led to the final quantum picture.) This magnetic effect can be used to make a spin detector. Arrange magnets as in Figure 4.8 to establish a definite direction along which to measure the angular momentum. Send a beam of atoms between the poles. A sensitive screen can register a dot where the emerging atoms strike. Spins pointing one way along the axis would be moved "up" toward a pole, opposite spins would be moved "down." Spins pointing at an angle would be in between. In this arrangement, first employed using silver atoms by O. Stern and W. Gerlach in 1922, the analyzer acts like a prism on a beam of spinning objects to create a *spectrum of spin directions*.[9] The classical spectrum would be a continuous "rainbow of directions" because the spins could be pointing any which way. As explained above, however, only a discrete set of dots is observed. What is surprising is that the spectrum of directions includes sets of *even* as well as odd numbers of dots.

For an isolated electron only *two* spin directions appear, "up" and "down" as in Figure 4.8(c). This is consistent with a value of n equal to $\frac{1}{2}$, so the magnitude of the electron "spin angular momentum" is $\frac{1}{2}(h/2\pi)$ and the allowed values of m are $+\frac{1}{2}$ and $-\frac{1}{2}$. It is this unexpected half-integral value that makes the spin peculiar. For electrons moving freely (no magnet), so there is no "up" or "down," the motion itself provides a natural direction for measuring spin orientation. Electrons spinning clockwise as you watch them moving away are said to be *right-handed* because the spin is in the direction of a right-hand screw advancing with the electron. *Up* and *down* are then replaced by the notions of *right-handed* and *left-handed*. Physicists use the Greek word *chiral* for "handed" objects.

Let me quickly correct the particle-centric language of the preceding paragraphs. Spin is a property that occurs at every point in space where the wave function does not vanish. Wherever an electron detector clicks, you could have chosen to detect its spin along an arbitrary direction that you define as "up," and the detector would *always* register either up or down along that direction. So there are *two probability waves*, $\psi\uparrow(x)$ and $\psi\downarrow(x)$, each of which will satisfy a different Schrödinger equation. The two cases are so similar that the two waves can be treated together in a spreadsheet-like approach as components of a single object called a *spinor*.[10] Then the two Schrödinger equations can be treated together as one "matrix equation" acting on the two-component spinor $\Psi = (\psi\uparrow(x), \psi\downarrow(x))$, just as the simultaneous equations in our matrix example above can be replaced by a simpler matrix notation. This is what I meant when I said that spin introduces more structure to Nature's geometry. It is always present, regardless of the spatial or temporal behavior of the wave function. The two-fold nature of the electron state can be revealed by a detector set to measure the spin components along an arbitrary direction.

4.6 SPIN IS NOT ENOUGH

Schrödinger's "law of motion," like Newton's, relates how a system changes in time to its configuration in space. For Newton, "configuration" meant positions and momenta of all the atoms. For Schrödinger, it meant the pattern of the wave function throughout space.[11] In both cases the equations of motion determine formulas for the energy of a system, and *vice versa*. Adding spin means adding a new piece to the energy that depends on the components of the two-fold spinor wave function, which in turn contributes a new term in Schrödinger's equation. The resulting equation accounted for the doublet lines in atomic spectra, but the predicted wavelengths were still slightly off the observed values. The atomic mechanics of the pre-Heisenberg/Schrödinger era had found *ad hoc* ways to patch up Bohr's old formulas to get the right values. And they involved taking into account the relativistic effects of the speeding electrons as they veered around the nucleus in the Bohr model.

Schrödinger had tried "relativizing" his equation, but its predictions were worse than the unpatched Bohr model, so he decided to postpone the problem and publish the nonrelativistic version first. Thus emerged the vexing question: "What is the correct relativistic form of Schrödinger's equation?" Schrödinger's straightforward approach had failed, and the increasing flood of empirical data seemed to offer no guidance for advancement. Dirac resolved the question in 1928 in a dramatic way that opened a new vista on Nature's landscape. He showed how to relativize Schrödinger's equation, but to do it he had to replace the two-fold spin wave function *with a four-fold array*. The resulting set of four simultaneous equations (or one equation expressed using four-by-four matrices operating on a four-fold spinor wave function) accurately reproduced the observed hydrogen spectrum and automatically included the electron spin energy with a correct coefficient that previously had to be adjusted by experiment.[12] In hindsight we can see how this result might have been anticipated (which I will explain in Chapter 6), but at the time Dirac's accomplishment seemed miraculous. Dirac himself confessed to have been "playing about with the equations" when he made his discovery.[13]

4.7 THE POSITRON INTRUDES

Each component of the two-fold spin wave function $(\psi\uparrow(x), \psi\downarrow(x))$ has a clear physical significance. The squared amplitude $|\psi\uparrow(x)|^2$, for example, is the probability of registering a spin-up event at position x. So what, if anything, do the other two components mean in Dirac's four-fold spinor $(\psi\uparrow, \psi\downarrow, \psi', \psi'')$?

In 1928, when Dirac relativized Schrödinger's equation, physicists thought they knew what everything was made of, namely atoms assembled from protons and electrons. The electrons, they believed, could be found either outside the atomic nucleus in Bohr's famous orbits, or inside where they cancelled the charge of some of the protons. Such cancellation would allow the nucleus to contain more protons than needed to balance the charges of the electrons *outside* the nucleus. Measurements had shown that atoms have masses equal to about twice the number of protons needed to neutralize the orbiting electronic charges. The proton-sized but electrically neutral *neutron* – which is *not* just an electron attached to a proton – was only discovered later in 1932 (see Chapter 6). How the nucleus could hold itself together against the tremendous electrostatic repulsion among the uncancelled proton charges was a puzzle, but no one was thinking about new particles. The ones that had already been discovered ought to have explained everything, if only they could be put together the right way.

So it was that Dirac himself failed at first to realize that his four-component wave function implied the existence of a new object identical to the electron but with opposite electric charge. He eventually declared, however (pressed by Hermann Weyl, among others), that the extra components described an entirely new feature of Nature – a *positively* charged electron (1931). An object with the requisite properties – called the *positron* – was shortly thereafter discovered in cosmic ray experiments, and Dirac's theory became the subject of awe. As did Dirac himself. Once again the effort to make the rules of Nature's dance independent of the observer's state of motion (through relativizing Schrödinger's equation for a spinning particle) led to the prediction of new phenomena – in this case a previously unsuspected new "particle."[14]

Here is the full picture: You can imagine objects like electrons but with different values of intrinsic spin. Corresponding to the electron's "two-ness" the other objects could have "three-ness", "four-ness", etc. with wave functions that are spinors with an appropriate number of components. Schrödinger's first effort to relativize his wave equation led to an object with multiplicity "one" (corresponding to no spin at all), which is why it gave the wrong results. But there could be other objects out there in the universe that might not be spinning, or that spin more than electrons. If you tell me what spin you need, I can produce a relativized Schrödinger/Dirac equation that describes it. In the limit when the average velocities of objects described by the equations are far below light speed (they have to have masses to go slower than light), *all* the different multi-component equations reduce approximately to the same simple nonrelativistic form, which is just the one Schrödinger first published. Dirac's work implies that the relativistic versions of all these equations require objects with a spectrum of discrete components equal to *twice* the spin multiplicity. Half the components refer to the original "particle," the other half to what we now call its *anti-particle*. This will be true for every fundamental object, not just electrons. Dirac in effect discovered not one new object, but a whole set of them. If you happen to find a new "particle," Dirac guarantees that you will also find its "anti-particle." (Rarely, a neutral object may be its own anti-object.)

4.8 ANTI-MATTER

Among the neologisms coined by physicists, *anti-matter* has deeply stirred the popular imagination. In all respects apparently like ordinary matter, electrically charged "anti-particles" combine spontaneously with their ordinary counterparts and disappear. They leave behind radiation whose energy equals the relativistic $E = 2 \times mc^2$ of the original combined masses. An electron and a positron, for example, annihilate to produce oppositely directed gamma rays, each of energy mc^2 – about half a million electron volts and easily detectable. These are exploited in *positron emission tomography* (PET) scanning, a form of medical imaging in which artificially radioactive nuclei are attached to molecules – usually a kind of sugar – that are taken up by metabolically active cells in an organism. There the nucleus decays to produce a positron that annihilates with a nearby electron to make the gamma rays, which are detected externally, establishing a line upon which the cell must be located. The pattern of radiation from many such decays maps out the location of the active cells.[15] PET scanners are often located near *cyclotrons*, commercially available machines that produce fast protons that collide with targets to manufacture excited nuclei that emit positrons when they decay. Thus anti-matter sources are created routinely for commercial purposes.

Anti-matter's exotic reputation comes from its propensity for spectacular annihilation with ordinary matter. But it is tied inextricably to ordinary matter. Dirac's theory shows that the positron is one incarnation of an underlying stuff that comes in four different forms, two of which have positive charge, and two negative. We call the negatively charged incarnations the two spin states of an electron. When all aspects of the stuff are energized in close proximity, which might be visualized as a bound electron–positron pair, the resulting physical system is unstable, just as an excited electron–proton pair (hydrogen atom) is unstable. In either case, the system decays to a more stable state that includes radiation which, Maxwell reminds us, is as real as the particles that produce it. When the hydrogen atom decays, its new state has the character of radiation plus a proton and an electron, the latter being trapped in a state of lower energy. When the system of electron plus positron (called *positronium*) decays completely, the new state consists of two quanta of radiation.

In an environment sufficiently hot and dense, as we think conditions were at the birth of our universe (see Chapter 8), the reverse process can occur in which photons of thermal radiation are annihilated to create electron–positron pairs (and matter/anti-matter pairs of other elementary objects). As the universe expanded and cooled in its early stages, at some point the omnipresent thermal radiation fell below the temperature needed to create the electron–positron mass, and annihilations began to outpace the creation of new pairs. Why *all* the pairs did not then subsequently annihilate leaving nothing but radiation is not yet understood. There seems to have been a slight imbalance in the matter/anti-matter interaction leaving just enough ordinary matter behind to make our universe. This is a clue to a mystery as yet unsolved.[16]

And here is another clue: Nature's force of electromagnetism acts differently on two of the four objects represented by the components of the electron spinor wave function (positive charge on two components vs. negative on the other two). Given that spin is not your everyday angular momentum, we should not assume we know how Nature treats *any* of the four components. Today we know that another force in Nature, the *weak force*, also operates differently on two of the four objects, but not the same two. Left-handed spin components are *weak charged*, right-handed components are not. More about this in Chapter 7.

The four-component spinor structure of the wave function is common to all elementary entities of the Standard Model (quarks and leptons in Figure 1.1), excluding only the substance of the force fields that act among them (the bosons in Figure 1.1). The "elementary particles" are otherwise distinguished by their different masses and by the different kinds of charge they carry. Symbols like $\psi(A)$ for the wave functions of these objects really mean the array of four complex numbers $(\psi\uparrow(A),\ \psi\downarrow(A),\ \psi\uparrow'(A),\ \psi\downarrow'(A))$, where the primes denote the anti-particle partners to the unprimed components.

NOTES

1. *Mathematical things.* W. R. Hamilton was among the first to create such new objects, first with a treatment of complex numbers as number pairs, and then with *quaternions*, the first noncommutative objects in mathematics. Many references exist on mathematics for popular audiences. For starters I recommend Davis and Hersh (1981) and Kline (1985).

2. *Complex numbers.* The graphical depiction of complex numbers as arrows in a plane was invented by Jean-Robert Argand (1768–1822) to visualize the square root of minus one, $\sqrt{-1}$, usually denoted by the symbol i. Since the square of a number (positive or negative) is always positive, i cannot have the same meaning as an ordinary, or *real*, number. Mathematicians invented complex numbers specifically to accommodate the square roots of negative numbers. They are used widely in modern mathematics and physics, and they play an important role in our story. The arrow in Figure 4.1 representing an arbitrary complex number $U = (A, \Phi)$ can obviously be

specified by its vertical and horizontal coordinates in Argand's plane. The coordinates are then called the *real part* and the *imaginary part* of the complex number, ReU and ImU. These are related to the amplitude A and phase Φ (Greek *Phi*) by the trigonometric ratios cosine and sine: Re$U = A$ cosΦ, and Im$U = A$ sinΦ. Then U may be written as the real part plus $\sqrt{-1}$ times the imaginary part, or

$$U = \mathrm{Re}U + i\,\mathrm{Im}U = A(\cos\Phi + i\sin\Phi) = A\exp i\Phi.$$

The last form employs the *exponential function* "exp" which is neatly related to the trigonometric ratios, as shown, using the imaginary element i. It is *defined* by the property $(\exp x)\cdot(\exp y) = \exp(x+y)$, which makes the multiplication rule for complex numbers explicit: $u\cdot v = (a\exp i\cdot\phi)\cdot(b\exp i\cdot\theta) = ab\exp i(\phi+\theta)$. The addition rule is most easily expressed using real and imaginary parts: $u + v = (\mathrm{Re}u + \mathrm{Re}v) + i\cdot(\mathrm{Im}u + \mathrm{Im}v)$, which just expresses the fact that the vertical and horizontal projections of the arrows for u and v add to make up the vertical and horizontal parts of the sum $u + v$. For much more on complex numbers, see Paul J. Nahin's attractive history (Nahin, 1998).

3. *Operations as objects.* The systematic treatment of symmetry operations is a branch of *group theory* (see below *Symmetries (2): groups*), for which the best introduction at the level of this text is Hermann Weyl's famous 1951 lecture series at Princeton (Weyl, 1952). This book also contains a masterful short description of the essential properties of vectors and coordinate systems, and their relation to geometrical transformations. Livio (2005) is an outstanding popular history of the development of group theory with brief references to the Standard Model. Regarding operations as objects to be combined with new operations to make new objects obviously creates a risk of confusion. The original "objects" (such as the square in the example of the text) upon which the new "operations as objects" act become superfluous, just as schoolchildren no longer need the actual objects counted by integers (marbles, bottle caps) once they grasp the abstract concept of number.

4. *Dirac's book of poetry.* Many physicists list Dirac's work among their all-time favorite books. The fourth edition is still in print (Dirac, 1958).

5. *One dimension, roughly speaking, for each choice of A.* In the ideal Euclidean geometrical space there is a different location A corresponding to each real number, which is far more than you need to specify a "nice" function like the wave functions of physics. "Real space" almost certainly departs from the topology of the real numbers at the smallest scale, and even if it did not, specifying the wave function at each of these points is overkill.

How close the points have to be to define a smooth function depends on the smallest scale over which the function varies, which presumably has some finite limit for physical quantities.

6. *Motion of the state vector in Hilbert space.* This picture of the winding state vector glosses over the subtlety that we freely choose *when* as well as *where* or *what* to measure. The Dirac view separates the abstract state vector from all of these contingent choices. Moreover, the principles of relativity demand that we treat *when* and *where* (time and space) on a more or less equal footing. Thus a better choice for an omniscient picture corresponding to the classical "tapestry of world-lines" is a Master State Vector for all of Nature in an abstract Hilbert space "outside of space and time" whose coordinates bear labels for time as well as for distance or momentum and other properties. In this picture space-time, being a property of the detectors, not the underlying stuff, is a macroscopic phenomenon. For historical reasons, the picture in which the state vector depends on time is called the *Schrödinger picture*, and the alternative static case is called the *Heisenberg picture*.

7. *Complementarity.* A notion introduced somewhat vaguely by Bohr at about the time Heisenberg proposed his uncertainty relation (1927). Bohr believed then that a complete description of nature required both "wave" and "particle" descriptions but since by the uncertainty principle they cannot be given simultaneously, he regarded them as "complementary." From the viewpoint of the Hilbert space description, the significance of complementarity is that a state vector can be represented completely in either the position or the momentum representation, but their associated coordinate frames do not coincide. I think of complementarity more generally as applying to *any* two frames of reference that do not coincide, not necessarily just those related to each other as are position and momentum.

8 *Spin.* In the decades prior to quantum mechanics, physicists had produced empirical formulas accounting for the multiplets, but failed to provide a physical basis. Wolfgang Pauli (1900–1958) first proposed a new "internal" degree of freedom to account for the two-valuedness (1925), but did not associate it with spin. G. Uhlenbeck and S. Goudsmit did, a few months after Pauli's proposal, despite the fact that to get the right magnetic moment, the electron would have to be spinning faster than the speed of light at its periphery (if it were modeled as a finite-sized sphere of "real" matter, which

it is not). Heisenberg thought the hypothesis was "brave." Pais (1986) gives a good account of the historical details.

9. *Magnet as a prism for spin.* *Electron* spins are actually not measured this way. A magnet will bend a stream of moving charges even if they do not possess a spin. This makes it difficult, for particles of very small mass like electrons, to separate the motion due to the spin magnetic moment from the motion due to the particle moving through the magnetic field. See chapter VI.1 of Wheeler and Zurek (1983). In practice, spins are measured on electrons attached to atoms whose neutrality removes this obscuring effect, first noted by Bohr.

10. *Spinor.* *Spinors* and *vectors* can both be represented by arrays, but the word *vector* is reserved for arrays whose elements act like coordinates in a space (e.g., spatial separation, velocity, and force are vectors). The components of a multi-component wave function do not refer to directions in space, and so a new word is needed to describe the underlying array. E. Cartan invented (discovered?) these mathematical objects in 1913, but according to Pais (1986), P. Ehrenfest first used the word "spinor" in a letter to B. van der Waerden asking if there were a "spinor analysis" analogous to "vector analysis" known to mathematics. The "weak force" in nature does not treat left- and right-handed spinor components of electrons identically (see Chapter 7).

11. *The pattern of the wave function.* The title (in English) of Schrödinger's first paper on wave mechanics is *Quantization as an Eigenvalue Problem* (Schrödinger, 1926). Schrödinger's equation, like Maxwell's, is of a kind (a *partial differential equation*) whose solutions are not numbers, but entire patterns, like waves, filling space and time, hence the name *wave equation.* Most solutions for waves that oscillate at a definite frequency (and therefore definite energy) are not physically reasonable – they approach infinite values at large distances. But for physical systems there is a set, a *spectrum*, of frequencies for which the solutions *are* well-behaved, and these are called the *eigenvalues* of the equation. Schrödinger showed that the energies of Bohr's allowed orbits are eigenvalues of the wave equation corresponding to a charged electron moving in the static electric field of the nucleus. Similar equations describe the ringing of a bell, and their eigenvalues are the frequencies of tones that sound when the bell is struck.

12. *Dirac's equation.* Weinberg gives a concise account of Dirac's motivation for seeking to replace Schrödinger's equation, and how he proceeded, in his

historical introduction to Weinberg (1995). The introduction is at about the same level as Pais (1986), which is written for a technical, but not expert, audience and which also includes a section on Dirac's equation. In case you wanted to know, here is what the Dirac equation for a free electron looks like in more or less modern notation:

$$\hbar\gamma^{\mu}\partial\psi/\partial x^{\mu} + mc\psi = 0.$$

The object ψ, as explained in the text, is a four-component array of complex fields, and x^{μ} stands for one of the set of four space-time coordinates: $x^0 = ct$, $x^1 = x$, $x^2 = y$, $x^3 = z$. Similarly, γ^{μ} stands for one of a set of four simple 4×4 matrices that operate on the array ψ. The entries of the tic-tac-toe arrays of the γ's are all either zero or ± 1 or $\pm i$ chosen to make the four coupled Schrödinger equations relativistically invariant. Here c is the speed of light, $\hbar$ is a shorthand for $h/2\pi$, and m is the mass associated with the object ψ. There is an implied sum in the first term where μ takes on the values 0, 1, 2, 3. This formula is an example of how the introduction of symbols for various mathematical objects can simplify what would otherwise be a complicated-looking set of four simultaneous equations, each with five terms. In practice, theorists usually employ even more abbreviated notation.

13. *Dirac's style.* "I think it's a peculiarity of myself that I like to play about with equations, just looking for beautiful mathematical relations which maybe don't have any physical meaning at all. Sometimes they do." Quoted by A. Pais in Pais *et al.* (1998).

14. *The positron.* Dirac at first associated the new spinor components with the *proton*. In the second edition of his famous monograph (Weyl, 1930), Weyl declared that Dirac's theory implied the mass of the proton should equal that of the electron, which is contrary to experience. The following May (1931), Dirac proposed "a new kind of particle, unknown to experimental physics, having the same mass and opposite charge to the electron." Later that same year Carl Anderson disclosed that he had found "anti-electron" tracks in cosmic rays (published in 1932). It appears that the experimental work was not motivated by Dirac's theory, but was part of a systematic study of cosmic ray phenomena. See Pais (1986) for details. Dirac's Nobel Prize was awarded in 1933, and Anderson's in 1936.

15. *Positron emitters.* Essential to the PET imaging process are nuclei that emit positrons in the first place. The first of these were discovered by the

husband–wife team of Irene Joliot-Curie and Frederic Joliot in 1933, shortly after the positron was discovered. The Joliot-Curies bombarded elements with fast helium nuclei (*alpha particles*) expelled in the natural radioactivity of polonium (discovered by Irene's famous mother Marie) to make artificially radioactive nuclei – for which they received a Nobel Prize in 1935. Their experiments were the first to create anti-matter in a laboratory.

16. *A mystery as yet unsolved.* The Standard Model allows for an asymmetry in the production of normal versus anti-matter, but one that is too small to account for the observed cosmic abundance of ordinary matter. See Chapter 7.

5   More is different

5.1 THE QUANTUM MICROSCOPIC WORLD VIEW: STEP 2
I promised an explanation of quantum theory in two steps. In the first
step Schrödinger's wave is something that moves in familiar three-
dimensional space. You could almost believe, as Schrödinger himself
did at first, that there is something real waving that corresponds to the
fundamental stuff of matter (electrons, quarks, photons, ...), something like Maxwell's dynamical field of electromagnetism that in the
old theory carries what we perceive as light. But alas, the Schrödinger
wave function is not even an ordinary number. Its symbol stands for a
whole array of numbers, including an *amplitude* at each point whose
square is the probability that a detector will click there, and components in a set of "internal dimensions" that are new features of Nature.
Worse, the Schrödinger wave for *two* detectors moves in *six* spatial
dimensions, three to locate each detector.[1] Adding more detectors
requires more dimensions. Whatever it is, the Schrödinger wave is
not a "real" field like Maxwell's.

Despite the unreality of the Schrödinger wave, its one-detector version $\psi(x)$ does very much resemble the electromagnetic field in Maxwell's
scheme. My accounts in later chapters of particle discovery and the
Standard Model are easier to think about if we imagine the one-detector
Schrödinger amplitude (squared) as not just a probability, but a measure of
an actual density of matter near a point in space. Most physicists indulge
this suspension of disbelief so they can apply their physical intuitions about
waves, laboriously acquired by solving problems about real waves of water,
light, or sound. Today we understand that *"real" waves are always macroscopic waves.* They entail huge numbers of Planck energy quanta. In
the microscopic domain only the probabilistic interpretation of the
Schrödinger wave is consistent with what we are able to perceive.

The second step of quantum theory sheds light on the connection between the unreal probability wave functions and the real waves we observe in Nature. By a clever and conceptually simple manipulation of the mathematical language, the immensely complicated system of multiple-detector probability waves in spaces of ever higher dimension can be related to a more intuitively accessible wave in ordinary space. Even so, the thing that waves is not something that can be measured in any conventional sense. Interpreting the wave functions for multiple detectors, moreover, discloses new phenomena that are non-intuitive in the extreme, and destroy once and for all any lingering hope that the microscopic world can be grasped with ordinary language.

5.2 SYSTEMS WITH MULTIPLE EXCITATIONS

Isaac Newton pictured Nature as swarms of Democritan atoms, interacting through forces similar to gravity. Omniscience meant knowing where each atom is all the time, which leads to the tapestry of world-lines I described in Chapter 2. Maxwell gave substance to the forces, adding a new but still visualizable complexity to the overall conception. Einstein modified the ideas of space and time that give meaning to the entire picture, a picture that physicists call *classical*, i.e., pre-quantum. It works well for big things, and any picture of tiny things had better merge into it and explain its accuracy. Be clear, however, that the larger view derives from the smaller, and not *vice versa*. How it does so is not straightforward. In the small, we focus myopically upon events of perception or *registration*, and space and time are not used to track particles but to locate the place and instant where such events may occur. The grand vision of the universe as an intricate tapestry of world-lines whose pattern might be assessed in one divine glance fails utterly in the microscopic realm. What replaces it is no less grand, but *different*. Among other things, it replaces the multiplicity of world-lines with a multiplicity of registration events. It is not far wrong to think of the large-scale world as a mega-detector whose visible contours register invisible behavior at the smaller scale.

Reality is stable because the components of the Standard Model (some of them) have clustered in our era in long-lived clumps: the stable nuclei of atoms and all the things they make. The clumps, as it were, are detectors for the microscopic events that brought them into existence. ("Dark matter," whose gravitational effects are evident in astronomical observations, is apparently also stable, and also clusters through gravitational attraction. Dark matter is important because it influenced the formation of galaxies when our universe was young, but it has not yet been integrated into the Standard Model.)

I tediously reiterate that the Copenhagen interpretation of quantum mechanics entails probabilities (squared amplitudes of wave functions) that depend on a set of numbers (location, spin, charges, . . .) that specify detectors at which events may be registered with the corresponding probabilities – "may" because we can choose to set an actual detector or not. The wave function gives the probabilities of events were we to arrange to detect them. How a particular setup of detectors responds to its environment depends on what is "out there." In quantum mechanics a "physical system" includes the detectors as well as the environment, and the two are linked in the wave function. Attempts to distinguish environment from detectors create conceptual difficulties that are intrinsic to the Copenhagen interpretation, as discussed in Chapter 3. One thing is clear, however: organisms such as ourselves are macroscopic, and so are the things we perceive as real, including the space-time framework we use to locate the large objects we call detectors. The wave function specifically links "something microscopic out there" to the macroscopic objects of our human-scale reality.

To be concrete, suppose "electron/positrons are out there." I think of a pervasive microscopic electron stuff that could be excited in different ways. By convention, I say that a certain environment has "one electron" if I set up lots of electron detectors but on repeated trials in the same environment, however I tune the detectors, only one clicks. "Two electrons" means two detectors click simultaneously, but no more, and so forth. I do not count "electrons," I count clicking detectors. If I have limited information about what is "out there," such as how much energy is available, I do not know in advance how many detectors will actually click, and I do not think of the energized electron stuff as possessing a definite number of "particles." A system having the charge of one electron could actually consist of one electron plus any number of electron–positron pairs. On some trials in a given environment one detector might click, and on other trials two or more might click simultaneously. For example, the energy in the environment could manifest itself as a click in a single detector tuned to register very high momentum, or as simultaneous clicks in several detectors tuned to lower momenta. How often either happens depends on the rest of the apparatus that produced the electron stuff.

Schrödinger's wave amplitude (squared) gives the probability for registering these various possible events. The event could be a click in a detector at location A, or it could be simultaneous clicks at locations A and B, or at A, B, and C. Thinking about multiple clicks takes us to Step 2 of the theory. If the machinery of quantum theory is to work for any kind of experiment, it must produce a probability for clicks in any number of detectors that we imagine. This requirement has profound consequences.

5.3 QUANTUM FIELD THEORY

Because we can imagine setting up huge numbers of detectors, and because the stuff of Nature can be energized as much as we like to set them all clicking, our conceptual machinery must be able to accommodate wave functions that depend on an arbitrarily large number of locations. This poses a tremendous challenge to our powers of visualization, and complicates the technical management of the theory.[2] I am going to tell you how physicists deal with this problem, not because it gives any new insight into Nature, but because it strongly shapes the appearance of modern physical theory and how physicists think about "what is out there."

Dirac was among the first to appreciate the need for a technical way to handle the proliferation of detector locations in the wave functions (he would have called them "particle" locations). I do not mean in the following to recount history, but to describe the logical position of Dirac's approach from a modern perspective. Recall that the wave functions for different numbers of detector locations: $\psi(A)$, $\psi(A,B)$, $\psi(A,B,C)$, ... can be written using Dirac's notation as $\langle A|\psi \rangle$, $\langle A,B|\psi \rangle$, $\langle A,B,C|\psi \rangle$, ... The quantum (Hilbert) space "coordinate" vectors $|A\rangle$, $|B,A\rangle$, $|C,B,A\rangle$, etc. are determined entirely by the detector setup. (The reversed label order in the "flipped" version of these vectors is a useful convention.) Now let us play the mathematical game of operators and objects (Chapter 4) with these multiple detector vectors as objects. Imagine operators that would have the effect of adding or removing a detector label. A *creation operator* (call it $\mathbf{C}^+(B)$ to create a detector at B) might operate on the vector $|A\rangle$ that depends on the single location A to make a new vector $|B,A\rangle$. This has nothing to do with experiments, it is a pure game we play with the detector state vectors. Write the result as $|B,A\rangle = \mathbf{C}^+(B) \cdot |A\rangle$. (Read it as "*vector B,A equals C-plus at B applied to vector A.*") The bold font reminds us that $\mathbf{C}^+(B)$ is an operator, not a number, and the dot reminds us that we are not multiplying here but operating somehow on $|A\rangle$ to make $|B,A\rangle$. An *annihilation operator*, which we can call $\mathbf{C}^-(B)$, would do the reverse so $\mathbf{C}^-(B) \cdot |B,A\rangle = |A\rangle$. *Define* a vector for the situation with no detectors anywhere, $|0\rangle$ say, and define the result of an annihilation operation on this to be zero, so $\mathbf{C}^-(B) \cdot |0\rangle = 0$. Using this set of mathematical things, and little more, you can rewrite the equations of the many-detector Schrödinger–Dirac theory so the number of detector locations in the wave function is handled automatically by the $\mathbf{C}$ operators, the only detector vector object that appears being $|0\rangle$. The good news is that the complicated many-location wave functions are pushed into the background. The bad news is that now the important objects of the theory are the *operators* $\mathbf{C}^+(A)$, $\mathbf{C}^-(A)$, etc., rather than the cumbersome but physically meaningful wave functions. In particular, the sequence $\mathbf{C}^+\mathbf{C}^-$ can act differently from $\mathbf{C}^-\mathbf{C}^+$. Mathematicians say the two operators do not *commute*.

This news is not all bad, however, because each creation or annihilation operator depends upon only a *single* location, and therefore it somewhat resembles a traditional wave function or field in three-dimensional space. To be sure, it is not a *number* that might give the magnitude of the field at that point, it is an *operator* with purely formal significance. We may call it a *field operator* or a *quantized field*, and when quantum mechanics is cast in this form it is called *quantum field theory*. The Schrödinger equation for the wave functions implies corresponding equations for the new field operators. *These equations for the operators strongly resemble the equations of classical physics for macroscopic field entities.* For example, the operator that adds a location to the detector vector for photons obeys an equation equivalent to Maxwell's equations. The corresponding operators for electron detectors obey the single-detector (or "single-particle") Schrödinger–Dirac equation. This is already evident from the relation $\psi(A) = \langle A|\psi\rangle = (\langle 0|\cdot\mathbf{C}^+(A))|\psi\rangle$, where the operator acts *backwards* on the flipped zero-detector coordinate vector. When A stands for location x, you can see that the wave pattern of $\psi(x)$ must be reflected in the operator function $\mathbf{C}^+(x)$.[3]

Also useful is the fact that the creation operators generate states in Hilbert space that do not have to be interpreted as defining detector setups. These states can also be interpreted as states of Nature that will definitely trigger detectors with given characteristics A, B, C ... And these are precisely the states experimenters attempt to make in the apparatus they use to probe their chosen targets (lasers, particle accelerators, electron beams, ...). See the Appendix on *How quantum mechanics is used.*

5.4 GUESSING EQUATIONS OF MOTION
Reducing the equations of motion – the Rules of the Dance – for multi-location wave functions to simpler rules for field operators is certainly useful, but where do the rules for these complicated wave functions come from in the first place? It took a Dirac to invent a satisfactory equation for the (single-detector) wave function of an isolated electron.

How do we guess the elaborate set of equations for the wave functions that have multiple detector locations for electrons, photons, and all the other Standard Model pieces? This is a task for which the field operators are particularly well-suited. Rather than guess at the full (infinite!) set of multi-detector equations for wave functions and then infer equations for the operators, do it the other way around. Use the fact that the equations for the operators are simpler, and guess *them*, then use the operators to construct the complicated wave functions and the formulas they imply for detector probabilities. These can be compared with experiments to check whether you guessed correctly. This approach shows the usefulness of purely mathematical entities in the process of discovery. The field operators are somewhat like the construction lines we draw to assist the proof of a geometrical theorem. They do not have direct physical significance themselves, but they vastly simplify the work.

For example, you could start with Dirac's one-detector equation and replace the simple one-detector wave function by its creation operator. Next use the operators to generate multiple-detector wave functions, and hope that the resulting probabilities correctly predict the statistics of detection events you actually see in experiments. Many technical puzzles emerge as you carry out this program mathematically, but all can be resolved (or worked around) to achieve a workable machinery for calculations.

This procedure, first developed by Heisenberg's contemporary Pascual Jordan, is usually called *second quantization* because the starting equations, Dirac's or Schrödinger's, for example, already describe quantum effects (like the hydrogen spectrum) for whose description only the single-detector wave function is required. Single-detector equations with a specified symmetry are (relatively) easier to invent (see Chapter 6), so this has been an enormously fruitful approach for discovering quantum equations of motion, but it is not without its problems. It leads to artifacts and inconsistencies in the second quantized equations (the equations for the field operators $\mathbf{C}$) that must be interpreted physically or removed "by hand" and it exaggerates the importance of the quantum fields relative to the more fundamental and physically meaningful wave functions.[4] It is true that in applications physicists hardly ever use the wave functions anyway but manage the technical apparatus to produce directly the probabilities of interesting events.

From Nature's point of view the process of second quantization puts the cart before the horse. The fundamental events for which the theory must account are detector registrations, and the fundamental objects of the theory are multi-component probability amplitudes for given detector setups (wave functions). The classical macroscopic field description of Nature emerges as an approximate consequence of the microscopic description in terms of probability amplitudes. Quantized fields are auxiliary entities introduced to expedite the technical management of the theory's mathematical structure.

Physicists sometimes say the microscopic objects of Nature "are" quantized fields. But "are" here implies a concept of reality, and we should be cautious in ascribing reality to something even more remote from experiment than the wave functions. The symbols for quantized fields refer to mathematical operators that create the coordinates of detectors in complex-valued wave functions whose amplitudes (squared) are probabilities for registrations in the detectors. The highly successful, but nevertheless approximate field equations that describe macroscopic reality – Maxwell's for example – are derived from and explained by the microscopic theory based on probability amplitudes, and not the other way around.

5.5 "STATISTICS"

Imagine an experiment with two detectors, one at A and another at B. Quantum theory gives us a single number – the squared amplitude of the wave function – that is the probability of hearing clicks in *both* detectors. Call this number P_{AB}. Since it makes no difference which detector is located in which place, this number must be the same as the number P_{BA} for the case where the detectors are interchanged. Therefore the wave function *amplitudes* must be the same for the two cases. But the *phases* do *not* have to be the same. On the other hand, the phases cannot be just anything. If the phase changes when you exchange the two detectors, it must return to its original value when you exchange them again. So either the phase is not affected at all by interchanging, or it is advanced by 180° (or retarded – either way gives the same result), which is equivalent to changing the sign of the wave function. In the second case interchanging again adds another 180° to make a full 360° cycle, taking the phase back to where it started. We call the two cases *symmetric* and *antisymmetric* under the operation of switching the detectors.

The two cases lead to strikingly different behaviors for matter that happens to be described by such wave functions. Suppose the two detectors approach each other so the probability for simultaneous detection approaches P_{AA}. Since an antisymmetric wave function changes its sign when detectors are interchanged, one arrangement approaches P_{AA} while the interchanged arrangement approaches $-P_{AA}$, which is impossible unless P_{AA} is zero. So *the probability of registering two events in the same place (or with the same detector tuning) is zero if they are described by an antisymmetric wave function.*

An object described by a *symmetric* wave function, however, can lead to clicks even if the detectors are arbitrarily close. If we decided to interpret the clicks as indicating the presence of particles (which is unwise), then two such "particles" could be in the same place at the same time.

Bosons and fermions. Both cases are observed in Nature. Leptons and quarks – the pieces of the Standard Model most like ordinary particles (the three left-most columns in Figure 1.1) – are described by antisymmetric wave functions whose phases change by half a cycle when the locations of two detectors are interchanged. These are the *fermions*. The remaining pieces carry various forces like electromagnetism among the fermions, and their excitations are described by wave functions that are unchanged when two detectors are switched. These are the *bosons*. The names honor Satyendra Nath Bose

(1894–1974) and Enrico Fermi (1901–54), who drew attention to the statistical properties of these objects. Objects like leptons and quarks that have spins equal to an odd multiple of the electron spin *must* have antisymmetric wave functions for the whole theory to be mathematically consistent. Objects with even multiples of electron spin *must* have symmetric wave functions.[5] Two fermion registrations can never occur simultaneously in the same place – a situation first appreciated by Wolfgang Pauli in whose honor it is known as the *Pauli exclusion principle*.[6] If two electrons (fermions) are captured in the vicinity of a nucleus, they can never be observed to have the same wave pattern (or "Bohr orbit" in the obsolete picture), unless they have opposite spins. Bosons, on the contrary, can pile up in arbitrarily large numbers at a single point or in a state corresponding to a single Schrödinger wave pattern.

Composite objects. Detectors can be designed to register composite objects like protons and neutrons, or nuclei, or atoms with their electrons, or even multi-atom molecules. The location labels $A,B,$ $C, \ldots$ might then refer to the center of a detector whose aperture is large enough to swallow the whole extended object. The quantum description remains valid, and a system composed of such objects must possess a wave function that is either symmetric or antisymmetric in the detector locations, depending on the overall spin of the whole object. Because of the short range of nuclear forces these objects are all localized and act very nearly like Democritan particles, which makes it reasonable to use the language of particles when discussing their properties. They are not quite like real particles, however – a point first appreciated by Einstein.[7] The probability of detecting two atoms will be described by a wave function $\Psi(A,B)$ with two lumps in its pattern centered at the atomic positions, say A_o and B_o. If these lumps merge into a feature comparable to the size of the detector aperture, which can happen as the atoms approach one another, or if they are moving very slowly so their de Broglie waves (the size of their lumps) spread widely, there is no way of telling which atom caused a click. Physicists say "the atoms are indistinguishable particles." Even in this case it would be better to say "the atoms are indistinguishable." Atoms do act like particles most of the time, but the probabilities of tracking (registering) them are given by wave functions whose arguments locate detectors, not particles.

Given the tendency either to avoid each other or to interpenetrate freely, you can imagine that collections of large numbers of either kind of object will behave differently. For example, the electrons (fermions) around a large nucleus, pile up in a ball whose size depends on the number present. Heavier atoms with more electrons tend to be larger (a tendency largely balanced by the stronger attractive electric force in heavy atoms). But if you crowd identical photons (bosons) into an optical box, such as between the mirrors of a laser, they will never overflow it no matter how many are present. The thermal properties of gases made of the two kinds of objects can be analyzed statistically, and they are said to obey *Fermi–Dirac statistics* (antisymmetric wave functions, fermions) or *Bose–Einstein statistics* (symmetric wave functions, bosons). The black body spectrum described in Chapter 3 is a typical statistical result for a "photon gas" in thermal equilibrium. Objects composed of odd numbers of fermions (protons + neutrons + electrons) are fermions, and those with even numbers are bosons. Except at very low temperatures the atoms of dilute gases have thermal velocities so great that their typical de Broglie wavelengths are much shorter than the distance between atoms, and quantum statistical effects are negligible. As the temperature is reduced below the point where de Broglie wavelengths of individual atoms overlap (i.e., peaks in the multi-detector wave function merge), the condition of symmetry on the wave function for bosonic atoms leads to a peculiar effect. The probability of registration is enhanced more than you would expect when a large number of detectors is tuned to the same state of lowest energy – the so-called *ground state*. That is, bosons tend to condense into the ground state at low temperatures. This is the so-called *Bose–Einstein condensation.*[7]

Why is the matter in the Standard Model fermionic, not bosonic? Perhaps for a different choice the resulting physics would not permit the kind of universe that produces physicists! Or perhaps there *is* a symmetry between bosons and fermions that is obscured by some mechanism that masks a simpler behavior of the pieces that interact with it (see Chapter 7). Nature would not have to choose between fermions and bosons if she chose *both*. So perhaps there are bosonic counterparts to each quark and lepton, but we just don't see them because they are too massive to be generated by the low-energy processes in our laboratories. The Fermi–Bose symmetry in this picture is called *supersymmetry*, and strictly speaking it is not (yet) part of the Standard Model.[8]

5.6 ABOUT DETECTORS

The preceding paragraphs illustrate how entirely quantum theory rests upon the notion of probability amplitudes – symbolized by $\psi(A,B,C,...)$ – for registrations in a hypothetical array of detectors tuned to the attributes $A,B,C,...$ When the detectors click, all we know is at that instant the part of Nature under inspection actually possesses the attributes for which the detectors were tuned. Thus we can learn from Nature only what we are clever enough to build detectors for. From this perspective, a principal aim of physics is to discover how many essentially different kinds of detectors we can make. The Standard Model is, among other things, an inventory of detector types. That Nature produces clicks in them gives evidence that this is Nature's inventory too. Nature is peculiarly intertwined with the detectors we build to query her.

Detector labels. The labels $A,B,C,...$, in the simplest case, denote points in space-time where detectors are, or could be placed, and each detector can register the presence of "something" there and then with a click. Usually we want to know what kind of something it is. That is, which piece of the Standard Model are we observing? Or which particular superposition of pieces? Different kinds of detectors respond to different pieces. So in addition to location, each label stands also for a whole set of detector parameters or tunings. Settings have to be declared for the various charges, for particle or antiparticle, and for internal spin magnitude and direction plus any other internal variables. Composite objects – atoms or nuclei, for example – will have even more parameters. Physicists rarely write these all down at once, but focus on just one or two properties of interest. The symbol "A" in $\psi(A)$ stands for the spreadsheet-like column of all the relevant detector settings needed to specify a unique object made from Standard Model parts.

These labels come in two varieties. One kind names the *choice of reference frame* for things like place (coordinate axes) or spin direction (direction of "up"). The other kind specifies the "tuning" within the range of possible results for that particular frame (values of coordinates x, y, z, or value of the spin projection number m). Electrons at a particular location, for example, can be in either of the two spin states $m = +\frac{1}{2}$ or $m = -\frac{1}{2}$. But these are registered with respect to some definite choice for the direction called "up." If we have only one spin detector that registers up and down along its own axis, and no other way of defining "spin up," there is no ambiguity. But two or more detectors may have their "up" axes aligned in different directions. So it is not enough to say "spin up." You have to say "up at 0°," or "up along an axis tilted at 5°," etc. Here the angle identifies the frame of reference, while "up" and "down" are the two possible outcomes within that frame. The angle itself, of course, must be measured in an arbitrarily selected "laboratory" frame in which all the detectors reside.

How measurement statistics depend upon this freedom to choose detector frames turns out to reveal yet another non-intuitive aspect of the quantum description of Nature. The interesting results appear in experiments with multiple detectors.

5.7 THE DISTURBING ARGUMENT OF EINSTEIN, PODOLSKY, AND ROSEN (EPR)

It is easy to imagine apparatus that will produce clicks in two separate detectors that are correlated in some way (I mean for the pattern of *clicks* to be correlated, not the detector settings, which are always under our control). For example, an excited nucleus might shed energy by radiating two electrons and nothing else.[9] Imagine that two

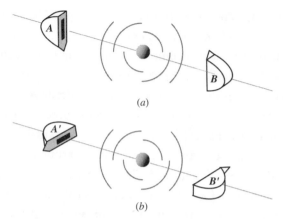

(a)

(b)

FIGURE 5.1 *EPR apparatus.* The excited spherically symmetric nucleus in the center serves as an apparatus to produce a two-electron state whose wave function has no preferred direction. Two detectors that can only register "spin up" along their axes are aligned opposite each other on a line through the center of the nucleus, with "up" directions opposed. In alignment (a), if detector A clicks then so will detector B. In alignment (b), if A' clicks then so will B'.

detectors, A and B, are set up to register them. If the original nucleus were standing still (zero momentum), then the electrons would necessarily be detected moving with identical speeds in opposite directions, because momentum is conserved even in the quantum world (equal and opposite momenta add to zero). So the detected momenta are perfectly correlated. If one detector clicks, the other will too. Similarly, if the excited nucleus had no spin then the observed spins of the emitted electrons must also cancel (to conserve angular momentum). If the spin detectors are both oriented along the same axis as in Figure 5.1 above, we can be sure this setup will cause one detector to register "down" if the other registers "up" *regardless of which direction is defined to be up.* Therefore the detector registrations in opposite settings at the two different detectors will also be perfectly correlated. Thus for this system *looking at detector A suffices to determine the result at detector B.* The microscopic electron stuff at A and B in this case is said to be *entangled,* a word coined by Schrödinger.

Einstein, Boris Podolsky, and Nathan Rosen (EPR) first drew attention to this two-detector phenomenon in 1935 to argue that quantum mechanics must be incomplete. They assumed the detection events are caused by two different particles, and that the particle at B or B' has an intrinsic spin that points in some definite direction. They could be sure of this because in either case its direction could be determined exactly by the registration at A or A' as explained above. That is, observing a click at A is essentially an experimental determination of the direction of spin at B. This satisfied a philosophical criterion EPR introduced for "an element of physical reality," namely, "*If, without in any way disturbing a system, we can predict with certainty ... the value of a physical quantity, then there exists an element of physical reality corresponding to this physical quantity.*"[10]

Now the detectors in Figure 5.1 could be very widely separated so, before its detection, the particle at B could receive no information about a registration at A except by a signal traveling faster than light. And yet the B particle seems to "know" that it must point its spin opposite to the A particle, regardless of how the experimenter orients the A detector. You might think the spin directions at A and B are fixed as the electrons leave the nucleus, but you have the option of turning A *while the electrons are in flight*. How does the nucleus know which way to launch the spin direction to match your choice of detector frame? The implication is that there must be further variables – *hidden variables* missing in the quantum description – that behave in some way as yet undiscovered to tell the particles how to behave. Thus quantum mechanics is incomplete.

"One could object to this conclusion," wrote EPR, "on the grounds that our criterion of reality is not sufficiently restrictive. Indeed, one would not arrive at our conclusion if one insisted that two or more physical quantities can be regarded as simultaneous elements of reality *only when they can be simultaneously measured or predicted* [i.e., with probability "one"]. On this point of view, since either one or the other, but not both simultaneously, of the quantities B and B' can be predicted, they are not simultaneously real. This makes the reality of B and B' depend upon the process of measurement carried out on the first system, which does not disturb the second system in any way. No reasonable definition of reality could be expected to permit this." (Einstein, Podolsky, and Rosen analyzed a different system, not a decaying nucleus, and I have changed the symbols in this quotation to match this case, first suggested by David Bohm.)

What is striking about the EPR essay is how purely philosophical it is, and how completely it hinges on an arbitrary definition of reality. Leon Rosenfeld, who wrote a commentary on this work in 1967, remarked that "Physical concepts, Einstein used to say, are 'free creations of the human mind.' In the case under debate, the 'criterion of reality' he proposed has very much this character, and it turns out to yield a striking illustration of the pitfalls to which one may be exposed by such arbitrary constructions of concepts. In spite of its apparent clarity, the criterion in question contains in fact a very essential ambiguity, hidden in the seemingly harmless restriction 'without disturbing the system.' To disclose this ambiguity, however, it is necessary to renounce any pretension to impose upon nature our own preconceived notion of what 'elements of reality' ought to be, and humbly take guidance, as Bohr exhorts us to do, in what we can learn from nature herself."[10]

The question of hidden variables. I looked seriously into the EPR "paradox" long after I was convinced that quantum mechanics is not about particles but about detection probabilities for experiments whose apparatus was an inextricable part of the system – which is consistent with Bohr's attitude toward the EPR objection. But I did not realize at the time that my confidence in this picture was premature. The brilliant mathematician John von Neumann had shown in a famous treatise that "hidden variables" of the sort needed for EPR's "complete theory" could not be incorporated consistently into the quantum framework (von Neumann, 1955). I labored over von Neumann's work in graduate school and came away convinced he was right. That David Bohm, a brilliant and unconventional thinker, had expressed doubts about von Neumann's proof in 1952 was unknown to me, as was the much stronger criticism of John S. Bell that appeared in 1965–6 while I was still a student. Then in

1969 I learned of a proposal by John F. Clauser for an experimental test based on a theorem of Bell's to determine whether hidden variables might indeed be possible. I was stunned to hear that some of the grand masters of theoretical physics, including Eugene Wigner and John Wheeler, believed that indeed the question of hidden variables remained open, and the proposed experiment would have fundamental importance.

5.8 BELL'S INEQUALITY

Consider what happens when the two detectors are *not* aligned. For example, suppose A is set at $0°$ ("up") and B is misaligned by $15°$ as B is in Figure 5.2. When the experiment is repeated many times, we expect to find clicks occasionally in the "wrong" channel – that is, A clicks and B doesn't, or *vice versa*. Call the expected frequency of "right" clicks in this case $R(0°,15°)$ and of "wrong" clicks $W(0°,15°)$, so $R(0°,15°) + W(0°,15°) = 1$ or 100%. We expect the same result if A is misaligned and B remains "down," because that just amounts to interchanging the detectors. So $R(-15°,0°) = R(0°,15°)$. In this notation, $R(0°,0°) = R$ $(15°,15°) = 1$ because aligned detectors always click "right."

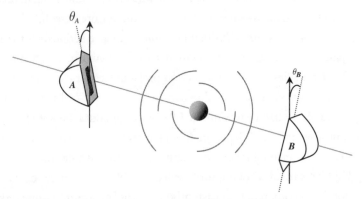

FIGURE 5.2 *Apparatus for testing Bell's inequality.* Simultaneous clicks in both detectors occur at a rate $R(\theta_A, \theta_B)$ depending on the tilt angles θ_A and θ_B. Bell's inequality, which assumes registrations at A and B are statistically independent, places an upper limit on the rate that differs from the predictions of quantum theory. The observed rates agree with the quantum prediction which exceeds Bell's upper limit for some range of angles.

In 1964, John S. Bell examined this kind of experiment more closely. Evidently tilting either detector may lead to wrong clicks. Suppose both *A and B* are tilted in opposite directions as in Figure 5.2 (so in our example the total misalignment would be twice as much, or 30°). If indeed the events at *A* and *B* are completely independent, as EPR would have it, then we have *two independent ways* of generating wrong clicks and therefore we might expect twice as many (i.e., $2W(0°,15°)$) compared to misaligning just one detector. Actually, less than twice because sometimes by chance both ways will operate so both detectors will click in the "wrong" channel. In that case, the clicks will still be in opposite channels, and we would observe this as a "right" rather than a "wrong" result. Call the frequency with which this happens $F(-15°,15°)$, which is a positive number less than one. Then $R(-15°,15°) = R(0°,0°) - 2W(15°,0°) + F(-15°,15°)$. Now notice that

$$R(0°,0°) + R(-15°,0°) + R(0°,15°) - R(-15°,15°)$$
$$= 2 - F(-15°,15°) \leq 2.$$

This formula, which is a form of *Bell's inequality*, is valid for any misalignment, not just 15°. It is important because it is precisely to preserve the independence of widely separated events that hidden variables would seem to be required. Bell's inequality gives a definite prediction based on the independence requirement that does not involve any quantum mechanical calculation of probabilities.[11]

It is not difficult (at least for a quantum mechanic) to write down the wave function for the two-electron excitations created when the nucleus decays, and use this to calculate the frequencies of clicks predicted by quantum theory for the experiment described above.[12] The result depends on the angle θ between the two detectors. Each term in the following formula corresponds to a term in Bell's inequality in the preceding paragraph:

$$1 + \cos^2 \tfrac{1}{2}\theta + \cos^2 \tfrac{1}{2}\theta - \cos^2 \theta.$$

For $\theta = (0°,15°,30°,45°,60°,75°,90°)$, this expression equals (2, 2.03, 2.12, 2.21, 2.25, 2.19, 2), which for a range of angles clearly exceeds the limit of 2 established by Bell's inequality. Actual experiments, carefully performed (using excited atoms emitting photons, not nuclei emitting electrons), *do* agree with quantum theory, so Bell's inequality is wrong.[13] This is surprising because Bell's analysis seems so obvious that it is hard to see what could be the matter. How can quantum theory violate this simple logic?

There is only one assumption in Bell's argument that could possibly be incorrect, and that is that the two ways of generating wrong clicks – tilting *A* or tilting *B* – are independent. Somehow the orientation of the detector at *A* must influence what happens at *B* and *vice versa* so as to introduce an additional correlation between events at the two detectors. This enhanced correlation is an essentially quantum phenomenon, and it is deeply disturbing. The *A* orientation can be changed just prior to the registration, so there is essentially no time for a signal propagating at the speed of light to reach from *A* to *B* before *B* clicks. Experiments show the extra quantum correlation even in this case, suggesting that the influence travels essentially instantly from *A* to *B*, that is, faster than the speed of light. Physicists call this kind of influence *nonlocal*. The hope of EPR for a "local" theory more complete than quantum mechanics apparently cannot be realized after all.

Of course we cannot identify a statistical correlation based upon a single measurement. An observer at *B* will see clicks randomly up and down, regardless of orientation. We have to collect data at *both A* and *B* and compare to establish the correlation, and this will take time. And no material influence is being propagated faster than light speed, so none of Einstein's strictures against faster than light propagation are violated.[14] What is very clear, however, is that these results are not compatible with any picture of Nature in which the object under inspection is isolated from the instruments of detection. *The correlations observed in the pattern of clicks in experiments with multiple detectors cannot be explained by a model in which the clicks are uniquely associated with stuff "out there" separate from the detectors.*

Subtle as the quantum correlation is, it establishes well-defined relations among separated events that might be useful in certain information processing applications. A whole new field of quantum information technology appears to be possible, at least in principle, based upon the evolution of specially prepared entangled states of light or matter within complex physical devices.[15] The technology requires, among other things, suspending the process of decoherence long enough to accomplish the desired information processing task. Over time the wave function describing the parts of the device involved in the task become hopelessly entangled with the much larger number of uncontrolled parts in the environment. Quantum information scientists work to delay this degradation or compensate for it with "error correction" schemes.

5.9 THE ENTANGLED UNIVERSE

Quantum weirdness appears where the external world joins our perception of it. And it appears most dramatically in apparatus with two or more detectors. By choosing what to look for at one, you seem to influence what happens at the other through what Einstein called "spooky action at a distance." The thorny issue here is not that measurement influences events, but that the influence seems to extend everywhere instantly and therefore "unphysically." Einstein had shown in his earliest work on relativity that the order of separated events is ambiguous if they follow each other too quickly to be connected by a light ray (see Note 29 in Chapter 2). This is the case with the correlated detector events in the EPR setup. Seen from different moving frames of reference a click at B could appear before, after, or at the same time as a click at A. In some frames the detector at B seems to "know in advance" how we would tune the detector at A. Even if we perceive this mysterious correlation through an autopsy of the data long after they are collected, it is still unsettling. It suggests a kind of pre-wiring of events at the microscopic level that vaguely conflicts with our notion of free will. How did B know to click exactly in the opposite channel of a detector at A even when we set the angle of A at the very last instant before hearing A's click? We can freely tilt A, but events at the distant B seem to know our choice immediately – even, in some relativistic sense, before we ourselves knew how we would set A.

The issue is much broader than merely a subtle correlation in a highly contrived experimental setup. Microscopic Nature does not need observers or their experiments to trigger a mouse-trapping registration. Everything we can survey with our gross instruments as "the real world" is an aggregation of Standard Model pieces that fell into place in a process similar to the act of measurement. Each large-scale object we examine can be regarded as a collection of Nature's own detectors. The vast number of particles, the randomness of their arrangement, and the complexity of their histories obscure any peculiarity that might have come from entanglement among the parts of the quantum wave function that described their genesis. But entanglement there must have been, given the origin of all material excitations in the strongly interacting environment of the Big Bang.[16] Mystics have expounded on this subtle connectedness among events, and compared it with an attitude toward reality found in some Asian cultural traditions. Nothing in those traditions, however, is comparable to the subtle machinery of quantum theory. Awareness of Asian philosophy does not help us understand the quantum behavior beneath reality, or *vice versa*.[17]

NOTES

1. *Schrödinger wave for multiple detectors.* All early interpretations of the spatial coordinate arguments of wave functions postulated the existence of a *particle* located at each position fixed by the coordinates. De Broglie confronted the problem of multiple particles already in his 1925 thesis (de Broglie, 1925), of which chapter 4 is entitled *Motion quantization with two charges*. After Schrödinger developed wave mechanics based on de Broglie's ideas, H. A. Lorentz said he preferred Schrödinger's account "because of its greater intuitive clarity [relative to Heisenberg's matrix mechanics], so long as one only has to deal with the three coordinates x, y, z. If, however, there are more degrees of freedom, then I cannot interpret the waves and vibrations physically..." Letter of May 27, 1926 in Klein (1967). I regard the coordinates as locating hypothetical *detectors* capable of registering a microscopic influence. This position avoids the very serious semantic difficulties of attributing a detector registration to a "particle." The more common usage of

physicists avoids these difficulties by redefining the word "particle," but I think this habit has created an unfortunate bias in the minds even of otherwise quite thoughtful scientists. See Note 19 in Chapter 3.

2. *Complications in the technical management of the theory.* Similar complications arise in any atomic theory of bulk matter, quantum or classical, where multiple particles interact with each other. The corresponding field of *many-body physics* therefore has a mathematical structure similar to quantum field theory, and technical methods for one are often useful for the other. The energy of a dense gas, for example, depends not only on the energies of its individual molecules, but also on the interaction energies within pairs, triplets, etc. If the molecules are not crowded too densely, then the extra contribution of larger clusters should be small, and the energy can be expressed mathematically as a sum of diminishing terms referring to ever larger clusters. Graphical methods similar to Feynman diagrams can be used to sort out the various contributions to the energy, giving a series of graphs similar to those in quantum perturbation theory. (See Note 15 in Chapter 7.)

3. *The history of quantum field theory.* Recounted in detail in Schweber (1994). The account is technical, but includes much biographical and historical information accessible to the devoted lay reader.

4. *Artifacts and inconsistencies introduced by second quantization.* One puzzle associated with second quantization is how to choose the order of the artificially introduced creation and annihilation operators in algebraic expressions. Creation before annihilation gives a different result than annihilation before creation – the operations do not *commute*. For example, the expression for the energy of the "vacuum" state $|0\rangle$ can lead to zero or an infinite value depending upon the choice of order. Since only energy *differences* are significant, physicists are comfortable choosing the order that gives zero. A deeper problem is the appearance of *anomalies* that violate a symmetry of the original equations. In the Standard Model, some anomalies introduced by second-quantizing the lepton fields and interactions are cancelled by anomalies from the quark sector. This can only happen when the number of quark families equals the number of lepton families, as in the Standard Model.

5. *The connection between spin and statistics.* Technically subtle. Feynman attempts a semi-popular exposition in his lecture *The Reason for Antiparticles* (Feynman and Weinberg, 1986).

6. *The Pauli exclusion principle.* Arrived in early 1925 – before quantum mechanics or even electron spin were discovered – to explain certain regularities in atomic spectra in the old quantum theory. Pais gives a complete account of this exciting period, with many references (Pais, 1986).

7. *Not quite like real particles.* In the summer of 1924, S. N. Bose communicated to Einstein a new way to derive the Planck distribution function, which today we recognize as equivalent to the condition that a "multi-photon" wave function should be symmetric under the interchange of "photon coordinates." Einstein pointed out that the same reasoning should apply to the waves associated with material particles, and discovered the possibility of condensation into the lowest energy state at low temperatures (Einstein, 1925). Hence the name: *Bose–Einstein condensation.*

8. *Supersymmetry.* As Dirac's relativization of the electron equation shows, choosing a symmetry that equations of motion must satisfy usually entails an increase in the number of components of the wave function. When one of the symmetry transformations converts boson-type wave functions into fermion-type wave functions, or *vice versa* (*supersymmetry*), an additional doubling of components is required. Just as relativity demands ordinary and anti-matter components for each fermion in the Standard Model, supersymmetry requires each elementary object to have fermion and boson components. The theory becomes technically complex because the entire group of transformations compatible with supersymmetry is large and its algebra is very rich. For a somewhat qualitative popular account, see Kane (2000). Woit (2006) has a more concise but readable account.

9. *Radiating two electrons.* This process is known as double beta decay. It has been observed (very rarely) when the decay is accompanied by two neutrinos, but never without neutrinos as contemplated here. If neutrinos were their own antiparticles, however, then the process could occur as described in the text (see Chapter 7). Several experiments are underway to search for neutrino-less double beta decay precisely to pin down the nature of the neutrino. Most EPR-type experiments are not done with electrons or atoms, but with photons generated by carefully excited electronic states in atoms.

10. *Difficult for Bohr and his colleagues to understand.* The EPR paper is reprinted in Wheeler and Zurek (1983), along with Leon Rosenfeld's 1967 commentary. Rosenfeld, who was working with Bohr at the time, wrote that during their attempt to analyze the EPR paper, "Now and then, he [Bohr] would turn to me: 'What *can* they mean? Do *you* understand it?'"

11. *Bell's inequality*. This version of Bell's inequality was first proposed by Clauser *et al.* (1969) and is called the CHSH inequality in the literature. An excellent and semi-popular account of another version of Bell's inequality was provided by Mermin (1985). Penrose (1989) (Note 14 in Chapter 7) reports a simpler analysis closer to my description which he also attributes to Mermin. Bell himself gives a popular account (Bell, 2004), where his original papers are also reprinted.

12. *Frequencies predicted by quantum theory*. For the record, the emitted electrons are in the singlet state $|\psi\rangle = \sqrt{\frac{1}{2}}(|{\uparrow}\rangle_A|{\downarrow}\rangle_B - |{\downarrow}\rangle_A|{\uparrow}\rangle_B)$, the state of the detector setup is $|D\rangle = (\cos\frac{1}{2}\theta_A|{\uparrow}\rangle_A + \sin\frac{1}{2}\theta_A|{\downarrow}\rangle_A)(-\sin\frac{1}{2}\theta_B|{\uparrow}\rangle_B + \cos\frac{1}{2}\theta_B|{\downarrow}\rangle_B)$, and the frequency of "right" clicks is $2\times|\langle D|\psi\rangle|^2$. The factor of two is for the two ways of getting "right" clicks: up at A, down at B and *vice versa*. Since $\langle{\uparrow}|{\downarrow}\rangle = 0$, $\langle{\uparrow}|{\uparrow}\rangle = \langle{\downarrow}|{\downarrow}\rangle = 1$, then $R(\theta_A, \theta_B) = (\cos\frac{1}{2}\theta_A\cos\frac{1}{2}\theta_B - \sin\frac{1}{2}\theta_A\sin\frac{1}{2}\theta_B)^2$. The half-angles are characteristic of spin-$\frac{1}{2}$ states which change sign under rotations of 360°.

13. *Bell's inequality is wrong*. No reliable observation has been reported favoring Bell's inequality over the quantum predictions. Experiments to date, however, have not completely excluded every conceivable possible loophole that might allow some kind of local hidden variable to exist. Early experimental papers are reprinted in Wheeler and Zurek (1983). Alain Aspect, whose group at Orsay, France made significant technical improvements in the measurements, reviews the experimental status in a chapter in Bertlmann and Zeilinger (2002).

14. *No material influence is being propagated faster than light speed*. Quantum mechanics itself can be used to show that quantum effects cannot lead to information transfer faster than light speed. The result is known as *Eberhard's Theorem* after P. H. Eberhard who demonstrated it in 1977.

15. *Quantum information technology*. Discussed in several of the essays in Bertlmann and Zeilinger (2002). The U.S. National Institute for Standards and Technology published a nice introductory booklet available on its Quantum Information Program website (http://qubit.nist.gov), or in hardcopy as NIST Special Publication 1042, 2005.

16. *Entanglement there must have been*. Consequences of this universal entanglement have been elaborated imaginatively by David Bohm (1917–92), whose concept of "implicate order" has attracted a following among nonphysicists and a great deal of skepticism by physicists (Bohm, 1980). In a late interview Bohm said "Initially I followed Bohr's view...

However, I later became dissatisfied with the general interpretation of the theory because it didn't give a clear concept of reality. It only discussed what could be observed and measured. If you say 'Fine, that's what it is,' then you would still raise the question: 'What can we nevertheless say about the nature of reality?'" Question: "What was reality for you, then?" Answer: "Well, reality would mean something that would have some existence independently of being known. It might be that we would know it, but it didn't require that we would know it in order to exist. Now it was difficult to see how this could be sustained finally in Bohr's view [i.e. the Copenhagen interpretation]. I proposed another model that had some interesting implications, but it was not received well. Essentially the leading physicists did not accept it. Later I came to the implicate order, which had a similar aim." Reprinted in Nichol (1998).

17. *Asian philosophy and quantum theory.* Bohm himself (see previous note) was drawn to prominent figures in contemporary Asian thought, including the Dalai Lama and particularly J. Krishnamurti, in search of a framework for a more satisfactory image of quantum reality. Capra (1991) has been a popular reference on parallels between quantum theory and Hinduism, Buddhism, and Taoism since it first appeared in 1975. Most physicists are uncomfortable with Capra's implication that there is more than coincidence in these parallels. The similarly popular Zukav (1979) refers to them in a somewhat less controversial way. The Buddhist point of view is expressed authoritatively by the Dalai Lama in a series of dialogues reprinted in Zajonc (2004). The dialogues are with well-known physicists who are experts in various fields, and the Dalai Lama responds to their accounts of quantum physics with perspicuous candor.

6 The machinery of particle discovery

J.J. Thomson's 1897 discovery of the electron was the third of three rapid-fire developments at the end of the nineteenth century that launched the campaign, still in progress, to understand the inner structure of atoms. The other two were complete surprises: *X-rays*, discovered by W.K. Roentgen in 1895, and *natural radioactivity*, discovered in uranium by H. Becquerel in 1896. Because the X-rays seemed to come from a glowing spot on the glass wall of Roentgen's apparatus, Becquerel immediately investigated whether phosphorescent materials could be induced to emit similar rays following exposure to the Sun. His discovery that uranium did indeed emit penetrating radiation, *without* benefit of sunlight, is surely the most profound purely accidental discovery in physics.[1] Until Einstein's discovery that mass is energy, it was also among the most puzzling: Where did the enormous radiant energy come from? (See Note 27 in Chapter 2.)

Each of these discoveries provided new experimental tools. X-rays, very short wavelength electromagnetic radiation, are a workhorse not only in medical imaging but also in the examination of microscopic structures comparable in scale to the X-radiation wavelength. Electron beams have even shorter de Broglie wavelengths that can be exploited in "electron microscopy" (among a multitude of other applications). The uranium emanations, however, later classified by Rutherford into alpha-, beta-, and gamma-species, so far exceeded the others in energy that they alone could serve as probes into the extraordinarily tiny and tightly knit structure of the nucleus. Within two years of Becquerel's discovery, Marie and Pierre Curie (who coined the word *radioactivity*) had discovered more radioactive elements, and particularly radium and polonium, whose energetic alpha emissions (bare helium nuclei) were indispensable to the birth of nuclear physics. Rutherford used them as bullets whose deflection in atomic collisions allowed him to determine the size of the nucleus (see Note 7 in Chapter 3). James Chadwick used them later in his 1932 discovery of neutrons, knocking them out from nuclei of beryllium. See Pais (1986) for an excellent brief history of this period.

Particle discovery, then as now, requires three things: a theoretical framework in which the notion of "particle" may be defined and its manifestations interpreted; a means of detecting the presence of objects so defined; and a way to induce these objects to appear at all. As to the first requirement, the concept of "particle" is subtle even in classical physics, implying indivisibility and a boundary between itself and a surrounding vacuum, among other things.[2] Maxwell's theory, quantum mechanics, and the properties of strong nuclear forces, required a new point of view that is part of our story. Detection, the second requirement, is a challenge to experimenters, but all particles possess by definition the property of interacting with other matter that in principle can be exploited to register visible traces (sometimes with great difficulty). It is the third requirement, finding or creating a source of particles, that presents the greatest challenge of all. Most of what we have decided to call "particles" are unstable and have to be manufactured "on the spot" by converting energy into mass. Focusing ever greater energies into progressively smaller volumes becomes increasingly expensive. So we nibble away at the low end of a possibly limitless range of massive objects. Who knows what Nature is hiding at energies beyond our reach?

6.1 MAXWELL'S IMPACT (3): ATOMISM UNDERMINED

That all things might be made of tiny parts is the leading reductionist paradigm in science. Like so many ancient ideas about Nature its realization at the smallest scales was deeply undermined by Maxwell's discovery that electromagnetic forces are "real stuff."

The problem is that pieces of charged matter and the forces among them are mixed inseparably together, and in Maxwell's theory the force field carries energy of its own. So how do we disentangle the properties of the source matter that carries the charge from the properties of the space-filling force field attached to it? An electrically charged object carries along an electric field. The field possesses energy. Energy is equivalent to mass. When we measure the mass of a charged particle, how much comes from the electric field energy, and how much from its "ordinary" mass? How do we tell which is which? You can measure mass by observing how an object accelerates under a known force. But the result lumps together the different contributions. Perhaps it makes sense to call the whole combination a "particle," but such a particle is nevertheless a composite of "matter" and "field." This is the third major awkwardness presented by Maxwell's theory. It did not cause a revolution comparable to special relativity or quantum theory (at least not yet), but it profoundly flavored the character of physical description.[3] Certainly the tale of "particle plus field" leaves much untold.

The Standard Model is a conceptual structure, a set of ideas, in which a small number of fundamental things appears in terms of which all else may be accounted for. But those fundamental conceptual things are not related simply to objects we can weigh or deflect in the laboratory. *Each "elementary particle" we actually observe is a combination of all the pieces of the Standard Model*, just as the observable bits of matter in Maxwell's theory are part charged matter and part electromagnetic field. Maxwell's field itself links all the electrically charged pieces of the Standard Model, and the remaining forces, "weak" and "strong," link in the rest. If in Nature the energy associated with a field were always much smaller than the energy $E = mc^2$ of its charged material source, then we would not be far wrong to associate the observable objects uniquely with each of the fundamental things in the Standard Model. Nature has not been so accommodating. Our ability to link "elementary particles" with Standard Model pieces depends on the strength of the forces they generate. I will revisit this problem in Chapter 7.

To make this idea of "linking" more concrete, imagine that all forces are turned off. Then the Standard Model would include only the 12 "elementary" fermions in the three left-hand columns of Figure 1.1 (this example ignores the three-fold nature of the quarks, described in Chapter 7). Considering only the "single particle" case (fields not quantized), each object would be described by a four-component wave function that evolves in time according to its own equation of motion, namely a Dirac-type equation with appropriate mass. That makes 12 Dirac equations, each of which could be solved by itself without regard to the others. When the forces are turned on, new *interaction terms* appear in each equation that are proportional to the charges each object carries and to the wave functions describing the other objects that have charges of the same type, so an equation for one object is now linked to all the other equations for similarly charged objects. The equations can no longer be solved separately, but must be solved all together. Easy to say, and easy for Nature which needs no mathematics, but our weak powers of analysis are not up to the task, and we must always resort to approximations. The quantized field trick at least allows us to handle the huge complexity of multiple argument wave functions that appear in Step 2 of the theory. This entire scheme, including the trick of creation and annihilation operators to handle the multiple-particle issue, was invented by Dirac in 1927 to describe electrons interacting with electromagnetic radiation.[4]

Let me pause here to explain how this scheme visualizes the radiation process. In our "simple" example we have 12 four-component Dirac wave functions (six leptons and six quarks) all linked together into one huge 48-component object. The 48 coupled equations of motion tell how this monster evolves in time. That's the bad news. The good news is that in practical cases nearly all these components can be ignored. For example, the initial monster Hilbert space state vector, of which all these coupled wave functions together define a single coordinate component, might correspond to an electron in an excited state of a hydrogen atom, so all but one of the immense number of components is zero. As time evolves, that initial component of the

Hilbert vector diminishes and the other components begin to grow (most of them very weakly), including one or more components corresponding to electron states of lower energy accompanied by an excitation of the radiation field (a photon). After a sufficient time the initial component will be practically zero, and the components corresponding to the lower energy electron state plus photon would carry all the weight. We query the status at any time by projecting the evolving state vector onto a particular direction in Hilbert space corresponding to a photon detector that we set up. The squared amplitude of the component along that direction (summed over all lower electron states with the right energy) is the probability that a photon has appeared with the properties our detector is looking for. This account has no "jumping electron" – the entire process is continuous. At some point the detector clicks and we know the atom has emitted a photon corresponding to the tuning of the detector. The mathematical model, even were it to include all the machinery of the detector, does not tell us when the click occurs.

Now, however, we are faced with defining what it was that "caused the click." The actual evolving state of Nature possesses non-vanishing components corresponding to every other part of the Standard Model (and anything else we may not have discovered yet), because they are all coupled together by the equations of motion. We set our detector to trap a creature whose Hilbert space directions we ourselves defined. (A typical detector triggers on a *range* of directions in Hilbert space.) A click discovers that Nature's state vector had components in those directions, but we have no evidence (from that single click) what other components might have been in the mix. Particle discovery needs a theoretical framework to suggest how to set the traps. Theory needs registered data to advance. The machinery of particle discovery includes software as well as hardware.

6.2 PARTICLE SPECTROSCOPY

Of all the things we might observe about radiation the most important is its energy. As Bohr already perceived when he launched atomic theory, the radiated energy advertises the structure of "allowed" states of the radiating system. These *eigenstates* are determined in turn by the internal forces operating on the electrons. Classical spectroscopy observes light radiated from electrons in chemical atoms energized by a spark, or heat, or other means. The electrons tumble down from one eigenstate to another, shedding excess energy along the way as photons, the energy quanta of electromagnetic force. "Lines" in the resulting spectrum correspond to energy differences among the relatively (but not absolutely) stable eigenstates – the old Bohr orbits. The emitted quanta of light carry energies ranging from a fraction to thousands of electron volts (eV – the energy gained by one electron accelerated in the field produced by one volt, about 2/3 the voltage of a typical flashlight battery). Visible light quanta are in the 1 eV range.

The same concept of spectroscopy applies just as well to radiation from other things – nuclei, for example, whose constituents can be energized by swift beams from radioactive elements, or from cosmic rays, or accelerators, or nuclear reactors. At energies of thousands to millions of eV the resulting spectrum arises from decay of the excited states of moving protons and neutrons inside nuclei. This is the subject of *nuclear physics*. If the energy of the exciting beam exceeds about 1 GeV (= 1 billion or 10^9 eV), the emitted radiation includes new *"particles,"* often short-lived, that decay quickly away into lighter, more stable objects, including protons, electrons, and photons. During their brief lifetimes the radiated objects bear witness to their origin, and give up information about the physical processes that created them. This is *particle* or *high energy* physics.[5]

Thus at high energies the brightest features of the radiated spectrum, the "lines," are called *particles*. Today we think of the lines in optical spectra as comprised of *photons*, which suggests a particle-like aspect that strengthens the family resemblance of all spectroscopies. I would prefer to reverse the terminology, and call the "particles" lines as well. They are simply features – "spikes" – in the energy spectrum as in Figure 3.3. Multiplicities appear in this spectrum just as in the optical line spectrum. From the perspective of physical theory, spectroscopic work even at vastly different energies is all pretty much the same. The spectral lines reflect differences in the energy states within an excited target. From these data, theorists attempt to construct systems of equations for wave functions whose components account for the multiplicities and brightnesses of the lines. It is not a simple task because Nature's facts do not speak plainly for themselves.[6]

If the emitted energy is great enough, its particles may decay in more than one way, or *decay mode*. Repeated observations will sometimes see one kind of decay, sometimes another. A large number of experiments must be performed to find the relative frequencies or *branching ratios* of the different modes, and also to determine the average lifetime, or *half-life*, of each decaying object. These extra features of particle spectroscopy add complexity to the task of finding the appropriate equations to describe the underlying objects, but they also give more information to help choose among the possibilities.

Notice the distinction here between observed features – "lines" – in the energy spectrum and components of the underlying theoretical model. The lines are what we experience as "particles." We model their properties with a theory that simulates the experimental objects with combinations of its own distinct fundamental parts.

6.3 THE BIG MACHINES

Dirac's discovery of the positron sanctioned, as it were, the manufacture of new particles. Once again, the history of discovery is too rich to reproduce here. It began in the decade prior to World War II with observations on cosmic rays and radioactive emissions, and intensified after the war with the enormous particle accelerators made possible by war-time radar technology.[7] Briefly, accelerators endow charged objects such as electrons or protons with huge kinetic energies that are released in collisions with targets in a flood of new objects. According to $E = mc^2$, the masses of the objects produced this way cannot exceed the collision energy (divided by c^2). To see objects with more mass you need larger accelerators. During the third quarter of the twentieth century, each new accelerator disclosed many new micro-scopic objects. Because they were so large, the accelerators had to be funded by national governments, and so *Big Science* was born. Nations that could afford it competed for the prestige of fundamental discovery and – mindful of the lessons of the recent war – perhaps out of concern that awesome military power might once again emerge from micro-scopic discoveries.[8]

The sizes of the machines grew with the number of acceler-ation stages (in linear machines, or *linacs*), or with the number of magnets required to bend the trajectories in a circle to pass repeat-edly through a single accelerating stage (*synchrotrons*). The particle concept is a good approximation here because the quantum wave-lengths are small compared with sizes in the accelerators, and we can speak safely of trajectories. Particles with higher energy are harder to deflect, so the circular trajectories must be large – the largest today, the Large Hadron Collider (LHC) at the European Center for Nuclear Research (CERN) is 27 km around. Bunches of particles speed through the evacuated chambers of the machines, urged on by rapidly alternating electric fields in resonators ener-gized by circuits similar to those that power radar beams and phased to give each bunch a kick each time it passes through a resonator. The energy that can be delivered this way is limited only by the size of the machine, increasing according to a relativistic formula even as the speed of the bunches approaches the limiting light speed. Stronger magnets make the circular machines smaller, which is why the highest energy devices use superconducting electromag-nets. This strategy eventually fails for electron machines because electrons, compared with the much more massive protons, radiate away their energy too easily when forced to turn in an arc. The highest energy electron machines are therefore linacs. Accelerators were built in the decade after the war in the USA, Europe, and the Soviet Union.[9] And they inundated the world of physics with a bewildering variety of new microscopic objects. See the nicely illus-trated large-format history (Close *et al.*, 1987).

Not until the 1970s did the dust begin to settle. Since my objective is not to tell history, but to divulge our understanding of Nature, I will describe the results of the glorious quarter-century of particle discovery with little comment. Particle physicists during that period performed the same service to science as the nineteenth-century atomic spectroscopists did. They imparted energy to matter and examined the spectrum of the ensuing radiation.

6.4 NEUTRONS AND NEUTRINOS

Charged objects are easy to detect because their passage through normal matter creates a wake of electrical disturbance. Electrically neutral objects can leave traces too if they have other kinds of charge. Neutrons, for example, have ingredients that possess the "strong nuclear charge" that leads to significant energy transfer in collisions with nuclei. Neutrinos, however, only carry the "weak nuclear charge," rendering them inordinately difficult to detect. Which makes it all the more surprising that the existence of the neutrino was surmised before the discovery of the neutron. Weinberg (2003) gives a good brief history of these discoveries.

Neutrinos and Fermi's theory of beta-decay. Already in 1914 James Chadwick had shown that, unlike other radioactive emissions, the beta (electron) radiation did not appear in "lines" but in a broad smeared-out spectrum. Every effort to explain this phenomenon failed until Wolfgang Pauli proposed that a very light neutral particle – later called the *neutrino* – might emerge together with the electron and share the total energy available. The broad spectrum simply represented the continuum of different fractions of energy carried by each, the sum of the two energies being the same for each possibility. This hypothesis accounts for the observed electron spectrum, but the neutrino itself was not detected until 1955 (see Note 5 in Chapter 7). I mention this development here because it moved Enrico Fermi to propose a theory of beta-decay modeled after Dirac's scheme to describe electromagnetic radiation. The important difference is that in beta decay a nucleus passes from a higher to a lower energy state emitting *two* objects (electron and neutrino), while the similar process for an atomic electron emits only *one* (a photon). This difference requires a new kind of interaction term in the linked equations of motion that describe all the participants in the decay process (later replaced by one similar to the electromagnetic interaction, as explained in Chapter 7). Fermi's theory was the first application of quantum field theory to nuclear processes.

Neutrons. More important for our immediate purpose is the discovery of the neutron. Before Chadwick's definitive experiments in 1932 it was known that some form of highly penetrating radiation resulted from collisions of alpha particles with beryllium, the fourth lightest element. In this context "highly penetrating" means the radiation had no electric charge – Irene and Frederic Joliot-Curie guessed it was gamma radiation, energetic enough to knock protons out of "hydrogenous material" like paraffin. Chadwick determined that only a neutral particle with mass close to the proton's would transfer its energy with such effectiveness to hydrogen and other light nuclei. He named it *neutron* after an earlier idea of Rutherford's that nuclei should contain neutral entities composed of an electron tightly bound to a proton. But the spin and statistical properties of such a compound object contradicted other experimental evidence. The neutron is *not* just a proton–electron combination, but rather a spin-½ object like the proton, and with a very similar mass.

6.5 MORE INTERNAL "DIMENSIONS"? ISOSPIN

I have shaped this account deliberately to prepare you for the idea that the multiplicities discovered in the particle spectrum require new geometrical features to describe microscopic Nature. It was luck that the two-ness of optical spectra could be explained by something as visualizable as electron spin (a visualizability that may be misleading). Spin occurs in ordinary three-dimensional space. Dirac's work showed that the antiparticles too result from considering "spin," but in the *four* dimensions of space-time. (It is just a coincidence that the number of components of the electron wave function – four – equals the number of

space-time dimensions. Objects with more spin need more components.) So physicists did not need to expand their notion of Nature's
microscopic geometry to accommodate these new features. In the particle spectrum, however, new multiplets appear that suggest "motions"
that cannot be accommodated in the familiar space-time geometry of
Einstein and Minkowski, even when augmented by quantum phase.

The most striking example is the proton–neutron pair. Here are
two features in the spectrum of particle radiation with nearly identical
energies (their masses differ by about one part in a thousand) but
different electric charge. They are a *doublet* in the energy spectrum.
Why not think of them as two states of the same underlying thing, just
as electrons and positrons are, or the "up" and "down" spin states of
either? The probability of detecting a proton or a neutron could be
described by a two-component mathematical object (or rather one with
eight complex components because protons and neutrons have spins
and antiparticles and phases too, but suppress these for now to focus on
the new duality of proton/neutron). The mathematical description is
so similar to electron spin that in the 1930s physicists decided to copy
the whole spin formalism and apply it to this new duality in Nature,
regarding protons as "spin up" and neutrons as "spin down" *in some
previously undetected "space."* Physicists refer to this idea as *isospin,*
and to the new "space" as *isospin space.* It would be a hitherto unob-
served feature of Nature's geometry, in which objects could be "spinning" with different quantized multiplicities, just like real spin in real
space. Except that the new objects would not be described as "spinning" faster or slower. They would have a property analogous to spin in
the new space whose manifestation in our world of apparatus is interpreted as *electric charge.* Thus objects with different isospin would
have different charges.[10]

Isospin in atomic nuclei. This idea is more useful than you might
imagine. Atomic nuclei are just heaps of protons and neutrons, and
they are held together by a very strong force that does not seem to care
about electric charge. That is, there must be a kind of *non-electric
charge* in Nature that protons and neutrons possess. Protons strongly

attract protons and neutrons indiscriminately, and neutrons attract neutrons and protons just the same. To say it in our new language, *the nuclear binding force is independent of the direction of isospin in isospin space*. Now form a *total* isospin for the entire nucleus by adding up all the individual isospins of the protons and neutrons. The result could be anything from some large value down to zero, depending on whether the individual isospins are aligned all together or half "up," half "down" (the latter implying equal numbers of protons and neutrons). Whatever it is, the total isospin for a nucleus could point in different directions in isospin space. For each of the discrete directions allowed by quantum theory, we expect to find a nucleus with a different balance between protons and neutrons. Because the binding force does not care about the isospin direction, each of these nuclei should have about the same energy. That is, there should be *multiplets of nuclei*. The difference in energy among different nuclei in these multiplets should be accounted for mostly by the electrical charge difference between protons and neutrons, not by the strong nuclear force. Measurements on nuclei do show such multiplets.[11] So the concept of isospin is useful.

Is this enough to declare "isospin space" to be a new geometrical feature of the universe? Since we see no evidence of such a feature in large-scale phenomena, it must be somewhat like the new "dimension" of phase discovered by Schrödinger, mostly hidden from view in macroscopic processes. I will not keep you in suspense. Further discoveries showed that protons and neutrons are made of smaller components now called *quarks*, and the nuclear binding comes from a force, whose quanta are called *gluons*, that link the quarks together. The effectiveness of the isospin concept must find its explanation in the behavior of these new subnuclear objects. But in the pre-World War II era of discovery, no one knew about quarks and gluons.

Isospin is important for our story, but I must introduce yet another new idea that led to the prediction of yet another new particle but for a while impeded progress toward a deeper understanding of nuclear forces.

6.6 MESONS AND THE RANGE OF FORCES

Knowing that it comes from a deeper "quark" structure, you will not be surprised to learn that the binding force among protons and neutrons is complicated. But it does have a few simple properties. Great strength is one (compared with electric force, which it must overwhelm to confine the mutually repellant charged protons in a nucleus); indifference to electric charge is another. A third striking property is its very short range, associated with the tiny size of nuclei (compared with atoms). There is no evidence that the force holding chemical atoms together to make up molecules derives any of its strength from the nuclear force. Although there are weak but readily observable effects of nuclear magnetic moments, chemical binding forces come entirely from electromagnetism generated by the electrical charges on protons and electrons, and quantum effects such as the "quantum pressure" that keeps them from collapsing altogether (Chapter 3). This fact implies that the nuclear force dies away very quickly beyond the minute scale of the nucleus. By contrast, electrical and gravitational forces have very long ranges. Gravity, indeed, extends its reach across the vastness of the universe to rule the shape of space itself. The long-range nature of electricity is obscured by its very strength (compared with gravity). Naked electric charges quickly neutralize themselves by attracting oppositely charged matter. Gravity, by contrast, continually strengthens as matter accumulates around an attracting object.

Quantum theory gives us a simple way to think about the range of forces, and links range intriguingly to mass. The energy of two attracting electric charges consists of three parts: the mass-energy of each plus the energy in the intervening electric field. Suppose one of the attracting partners was not always present, but came into existence – perhaps by radioactive decay – at a certain time. Since the intervening field is not yet present, how does the other partner become aware of its existence? Where does the energy come from to establish the field? Energy, being proportional to frequency by Planck's relation, is not well-defined for brief processes (duration T), and by the energy–time version of the uncertainty principle additional energy E is available for a time inversely proportional to E: $T \approx h/E$. You can make electromagnetic excitations (photons) with arbitrarily small energies (frequencies), so E can be as small as you like. Therefore T can be arbitrarily long, giving plenty of time for the field to establish itself between distant charges. (See Note 22 in Chapter 3.)

Now suppose the photons have a mass m of their own (they wouldn't be called photons then!). The linking field must possess at least the energy mc^2. Such energy is only available for a short time interval $T \approx h/mc^2$, and in that time the field can travel no farther than $cT \approx h/mc$ (maximum speed c times travel time). Charged objects at greater separations than this have no time to "sense" each other. This gives a typical range for the force field. Thus *massive force fields imply short-range forces*. If you require the range to be as short as the radius R of a typical nucleus, you get an estimate for the mass of the force carrier: $m \approx h/Rc$. Hideki Yukawa, who first conceived this argument in 1934, therefore suggested looking for a new particle with this mass (a few hundred times the electron mass) among the debris of cosmic ray collisions. If found, it might confirm this explanation of the short range of the nuclear force. A new particle with about this mass *was* soon found, but it did not interact strongly with nuclei. Later, however, a second particle of the right mass appeared that did interact strongly, and everyone felt better (but uneasy about the first particle. As I. I. Rabi puzzled, "Who ordered *that*?").[12]

We know today that the puzzling first particle, now called the *muon*, is truly fundamental, closely similar to the electron, and completely unrelated to the strong nuclear force (as unrelated, that is, as any object can be from any other in a world of interacting quantum fields).[13] The second particle is one of a triplet known as *pions*, distinguished by their charges: plus, minus, and neutral (therefore an *isospin triplet* with symbols π^+, π^-, π^0). Today we regard each pion as made of two quarks (actually a quark and an antiquark), in contrast to the three quarks required to make a proton or a neutron. The massive pions are indeed closely related to the mediating strong nuclear force field, but once again things are complicated because they are not truly fundamental particles. They are an example of quark constructs called *mesons*. More about this in Chapter 7.

Yukawa's idea did not get to the bottom of the mystery of nuclear forces, but it was an important clue. The idea of a massive field of force was new. It further undermined the distinction between "matter" and "force," and encouraged a more abstract view of mass. But this basic view of nature is unchanged: matter comes in different kinds that mutually interact. Some kinds interact more strongly than others, and the strength is measured by a charge – different kinds of charge for different kinds of interaction. Mass is just a property – a threshold in the energy spectrum for each entity – that any of the kinds may have. Neither mass nor charge distinguishes "matter" from its linking forces. At its current, certainly incomplete, stage of development, however, the theory does treat "matter" differently from "force." The proposed extension of the Standard Model called "supersymmetry" puts them on a more equal footing, but no observations as yet require it to be a property of Nature.

6.7 IF ISOSPIN WERE "REAL"

Just as the Schrödinger field for a spinning electron has two components, there must be two components for the "iso-spinning" nucleon. Or rather, as I said before, *twice as many* components, because nucleons – protons and neutrons – also have antiparticles and ordinary spin, just like the electron. Mathematical notation has been invented to help keep track of all these components. The detail is incidental to the theme of our story, but necessary.

Remember what these components mean. We have detectors, or a single detector with filters in it that we can set depending on what kind of event we want to register. First set the location and time for the measurement. That tells us where and when to evaluate the Schrödinger/Dirac field components to predict the probability of the event. Next set the conventional spin orientation, say "spin-up at 0°," then set the matter/anti-matter filter for, say, "normal" matter, and finally set the nucleon filter for "isospin up" or protons. Then the squared amplitude of the field component – one of eight – singled out by these settings predicts the fraction of the time the detector will register a spin-up proton if we repeat the identical experiment a large number of times. We need to know what kind of experiment is being conducted in order to calculate the field in the first place. That information determines the initial conditions for solving the Schrödinger/Dirac equations.

Nuclear force explained? In the 1950s it occurred to Chen Ning Yang that if the nuclear binding force were truly indifferent to electric charge – that is, to isospin direction – then the orientation of the isospin axes (in the new hidden isospin space) *could be selected arbitrarily* at each point of normal space-time and the force should be unaffected. In a world where the nuclear force is the *only* force there would be no way to distinguish protons from neutrons at all, and the orientation of the isospin axes would truly have no consequences. We have seen this situation before with the unobservable Schrödinger phase. Yang and his collaborator Robert L. Mills viewed the new "internal" dimensions of isospin space as akin to the "dimension" of phase that is linked through Weyl's gauge theory to the electromagnetic field.

Weyl's idea, you will recall, extended Einstein's fruitful hypothesis that Nature's laws of motion should be independent of arbitrary observer choices. If we have the right theoretical description of Nature then Einstein believed that altering it by introducing different reference frames at different points in space should not lead to any differences in the predictions of the theory. When this principle is applied to Schrödinger's phase, which is part of the "space" in the quantum description, it can be made to work *only* if the theory contains the electromagnetic potential in a certain way. Then the pseudo-effects that would otherwise be introduced by the changes in the phase reference frame are cancelled by a certain adjustment to the potential (a *gauge transformation*) that does not have any physical consequences.

To ensure that the phase is truly unobservable, it was necessary to introduce the electromagnetic force via Weyl's gauge theory. Could it be that a similar theory can account for the nuclear forces? In 1954, while visiting Brookhaven National Laboratory, Yang and Mills worked out the analogue of Weyl's theory applied to the "isospin gauge," and discovered a new set of equations for the hypothetical nuclear force field analogous to Maxwell's equations. The *Yang–Mills equations* are more complicated than Maxwell's because they describe forces that can be "gauge transformed" to compensate for the arbitrary orientation of axes in three-dimensional isospin space, compared with the arbitrary origin of the phase angle (one dimension) that leads to Maxwell's equations.

I well recall my astonishment and delight when I first learned about the Yang–Mills theory of nuclear force in graduate school. It made an enormous impression on me. I was aware of the electromagnetic gauge theory, and thought it wonderful that Maxwell's equations followed logically from such a fundamental requirement as the unobservability of the quantum phase. That a similar idea should lead to the nuclear force seemed entirely natural. Unfortunately, the theory had a major drawback. It seemed to predict forces with *massless* quanta like photons, not the force whose massive carrier had been foreseen by Yukawa and presumably had already been discovered as the pion.

Wolfgang Pauli, a founding father of quantum physics, had anticipated Yang and Mills and had already explored an isospin-based gauge theory of nuclear forces. But he also knew about the mass problem, and decided not to publish. I credit Yang and Mills with the courage to declare the importance of their result despite this shortcoming. They responded to the beauty of the idea, and could not resist sharing it with others. Pauli once famously said of himself that he "knew too much" to choose productive physics problems after a certain age. Perhaps this was one of those occasions when his reluctance to advance a theory he knew to be wrong in a specific context prevented him from receiving credit for an idea so beautiful that it just had to be right somehow.[14]

For my own part, I learned of the Yang–Mills theory in the mid-1960s at the zenith of the particle discovery era. It seemed to me that particle theory was in chaos, and the most likely way out was in the systematic analysis of the particle spectrum using concepts that associated internal particle "dimensions" with the observed multiplicities. My first love, however, was not spectroscopy but field theory, of which the Yang–Mills result was a beautiful feature. Unfortunately, field theory did not then seem to be working very well in the particle world, and I turned to an area of physics where it might have more success – the new territory opened up by the invention of the laser in 1960. Much later I actually found a use for gauge-invariance ideas to solve a problem in nonlinear optics.[15] By that time, the Yang–Mills theory had borne fruit in particle physics, but in a totally unexpected way (unexpected by me at least).

At this point, I think I had better say something about "symmetries" and their relation to the particle spectroscopy that did eventually bring order out of chaos. Physics talk is full of references to symmetry that must be frustrating to the uninitiated. The significance of symmetry is not obvious, and an explanation, however cursory, seems to be in order.

6.8 SYMMETRIES (I): CONSERVATION LAWS

Newton's First Law of Motion is a *conservation law*: in the absence of forces, particles move in straight lines with constant speed. Speed and direction do not change with time: they are *conserved*. The Law has two parts: a statement of the condition under which something is conserved (no forces), and a statement of what is conserved (speed and direction). A more careful analysis shows that what is conserved is not just the speed but the *momentum*, which in simple cases is the product of the particle's inertial mass and its velocity. In physics language, the words *velocity* and *momentum* refer to direction as well as magnitude. So the First Law says *in the absence of forces, momentum is conserved*.

Think of what absence of forces implies. From the object's viewpoint, its path is clear. It will not have to work, i.e., to expend energy, to surmount obstacles along the way, or what is the same thing, its *potential energy* remains unchanged during the motion (this *defines* potential energy as the work required to go along a path). Thus the potential energy is indifferent to translations along the path: it has *translational symmetry*. So the first part of the conservation law is a statement about the symmetry of something. "Absence of forces" is identical to "potential energy is translationally symmetric." Physicists say "the *system* is translationally symmetric" because the motion of a self-contained system is completely determined by how its energy depends on the positions and velocities of its various parts. The other part of the energy, the *kinetic energy*, usually depends only on the velocities of components of the system, not on positions. Otherwise the entire energy, kinetic plus potential, must be translationally symmetric to conserve momentum. This way of depicting the energy – not as just a number associated with a particular state of a system, but as a function of the various possible states – is the basis for the formal mathematical treatment of equations of motion in physics.

Surprisingly, all conservation laws are like this. They state what is conserved if the system has a certain symmetry, the system being described by its energy expressed in terms of the things that could change as the system evolves. The kind of symmetry in question here is not like that of a square or a triangle, where a rather large motion is required to bring the object into a new position with the same appearance as the old. Here the energy (analogous to the "appearance") is unaffected by a translation of *any* size, large or small. This is a *continuous symmetry*. A picket fence also has a kind of translational symmetry, but you have to move the fence through one whole picket (and the gap next to it) before it looks as it did before the move. The fence has a *discrete translational symmetry*.

If you are sitting in a frictionless swiveling chair in the center of a round room with perfectly smooth, featureless walls, you cannot tell which direction you are facing. This is a situation of *rotational symmetry*.

As you turn around in the center, nothing catches at you to slow your motion. An object with no angular force on it does not care where it points. If it is set spinning with constant rotational speed, it will spin forever. That is, its spin, or angular momentum, is conserved. *Rotational symmetry about an axis implies conservation of angular momentum about that axis.* Another conservation law.

How about energy itself? A quiescent object can be kicked into motion at any time, increasing its energy, so energy obviously has not been conserved. To conserve energy, you must have a system that is isolated from external stimulation. The system has to be undisturbed "forever." You cannot tell from the behavior of the system what time it is. It remains the same under translations in time. *A system with time translational symmetry conserves energy.*

The universe (and therefore its energy) is indifferent to the particular value of the phase of Schrödinger's electron field, which might be called *phase symmetry*. What would happen if it were not? In this case the consequences are easiest to appreciate in the corresponding effect on the electromagnetic field, whose invariance under gauge transformations is implied by the phase symmetry according to Weyl's idea. If the fields are not gauge-invariant, then Maxwell's equations do not suffice to describe them, and in particular the relation between the fields and their sources (charges) implied by the equations fails to hold. Consequently the charge cannot be accounted for by following how it flows in response to the fields – and it will seem to change from place to place. That is, electric charge will not be conserved. *Gauge symmetry implies conservation of electric charge.* (The mathematical analysis is more direct than my verbal explanation, but presupposes too much special knowledge to admit a brief translation.)

I intend these rather casual remarks about the connection between symmetry and conserved quantities to give you a feeling for how the language is used, and what physicists mean by *symmetry* and by *conservation*. It is the *energy* of the whole system, or rather how the energy depends on different variables, whose symmetries are of interest here, and "conservation" implies constancy as time passes.

Recall that "energy" is a quantity *defined* by the equations of motion (Chapter 2), and conversely the equations of motion can be derived if the energy is known in terms of variables like positions and momenta. These equations determine how things move as time advances, so it is not too surprising that symmetry properties of the energy should be linked to different combinations of variables being conserved in time. This is true for quantum equations like Schrödinger's or Dirac's as well as for the "classical" equations of Newton or Einstein. From the perspective of a theorist, the whole scheme is a beautiful as well as a useful theorem of mathematics – a monument to Emmy Noether, who crafted it with elegance and perspicuity in 1915: *To every continuous symmetry of a system there corresponds a conserved quantity.*[16]

6.9 SYMMETRIES (2): GROUPS

Mathematicians have a way of looking at symmetry that was immensely valuable for physics during the golden era of particle discovery. It enabled physicists to discern symmetries in previously unsuspected dimensions by examining the pattern of multiplicities in the particle spectra. How this works is a fascinating story about the interaction between mathematics and physics that is going to be difficult to tell in a short space. But only a few basic points are needed to appreciate the idea. See Weyl (1952, note 6) for a nice introduction to these ideas.

Recall that the mathematical game of objects and operations finds sets of things like numbers, and invents operations on them like addition and multiplication. Of all the playful products of this game, the most incredibly fruitful has been to identify sets of operations *themselves* as objects. The natural operation like multiplication combining two "object" operations is simply to follow one by another. An example will make this clear, and will also demonstrate something surprising and deep about the ordinary process of turning things around.

Place a book on a table in reading position, Figure 6.1. Looking down, turn it clockwise about the vertical axis by 90° (Operation *A*). Next turn it counterclockwise 90° about an axis pointing to the right

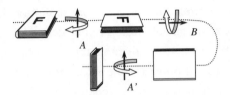

FIGURE 6.1A The sequence of rotations $A'BA$ gives spine *down*. (Letters symbolize rotations acting to the right, so read the order in which rotations act from right to left.)

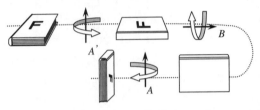

FIGURE 6.1B The sequence ABA' gives spine *up*, so $A'BA \neq ABA'$.

and parallel to the near edge of the table (Operation B). Finally turn it counterclockwise 90° about the axis of Operation A (Operation A'. The prime indicates A' is the *inverse* of Operation A). The result is the same as if you had turned the book up counterclockwise by 90° about an axis pointing away from you perpendicular to the edge of the table (Operation C). This gives meaning to the expression $C = A'BA$, where the letters refer to rotations, not numbers, and writing them next to each other is understood to imply a sort of multiplication – each operation acting to the right.

Now return the book to its starting position, and apply the operations described above, but in the reverse order. You will have to do it carefully, because it is tricky to keep axes and rotation directions straight. You will find that ABA' does not equal $A'BA$! (The spine now points *up* instead of *down*.) "Multiplying" two rotations gives a result that depends on which goes first. Mathematicians say the "multiplication" is *noncommutative*. This new property of "multiplication" leads, as you can imagine, to an "algebra" of rotations that is much richer than the algebra of letters that stand for ordinary numbers.

Mathematicians began to study the algebra of symmetry operations late in the nineteenth century. The collection of all symmetry operations of an object – operations that do not change its appearance – is called a *symmetry group*. The group contains the identity operation and the operations that undo or invert the other operations and all their products. Remarkably, it turns out to be possible to study the properties of groups even if you do not have a corresponding picture of an underlying symmetry. The groups of discrete operations, i.e., not continuous like the symmetries that give conservation laws, can be uniquely distinguished by their *multiplication tables*. Continuous groups can be distinguished in a similar way that I will describe shortly.

For example, an object with no symmetry at all (the letter "R" for example, or in three dimensions, a glove) has only one operation that brings it back to itself, namely the *identity operation* that does nothing at all. Its multiplication table is boring: essentially one times one equals one. Human bodies possess a single (approximate) symmetry of *reflection in a mirror plane* slicing us vertically into left and right. This reflection plus the identity operation together form a group (the only one) with *two* elements and another simple multiplication table (two successive reflections equals the identity). So there is only one group with just one operation, and only one group with two operations. And there is only one group with three operations (the symmetry rotations of the *triskele*, that three-legged figure associated with Sicily, or the insignia of the Isle of Man), but there are *two* possible multiplication tables for groups with four operations. It is not difficult to work out all the possible different multiplication tables for groups of higher order, but it does get tedious. Mathematicians have tricky ways of doing this that simplify the task.

In case you wanted to know, here are the multiplication tables for the only two groups with four elements (Figures 6.2a, b). Call them I, A, B, C, where "I" stands for the identity operation which does nothing. For example, in the "cyclic group of order four," $AC = I$, but in the "four-group," $AC = B$. For both these groups, multiplication is commutative. That is, $XY = YX$ for all elements. The cyclic group of order four describes the symmetry of a swastika, where A, B, and C are rotations by 90°, 180°, and 270°. The four-group describes the symmetry of a rectangular box with unequal sides that has been distorted so the angles on top and bottom faces are no longer 90°. A is then a rotation by 180° about the vertical axis, B is a reflection through the mid-plane, and C is *inversion* or reflection of every point on the box through its center. (In the tables below, the result of operating with an element in the left column on an element in the top row is in the cell at the intersection of the row and the column labeled by the two elements.)

Here is the table for a group with six elements where multiplication is *not* commutative (Figure 6.3). For example, $BD = E$, but $DB = C$ according to the table. This is the symmetry group of an equilateral triangle, A, B being rotations through 120° and 240°, and C, D, E being reflections in the bisectors of the vertex angles. Notice in all cases every element of the group appears once in each row and each column.

Continuous groups. The groups of importance for conservation laws – *continuous groups* – each have an *infinite* number of

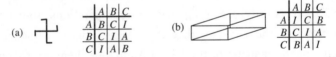

(a)

	A	B	C
A	B	C	I
B	C	I	A
C	I	A	B

(b)

	A	B	C
A	I	C	B
B	C	I	A
C	B	A	I

FIGURE 6.2 Multiplication tables for (a) the cyclic group of order four; (b) the four-group. The figures next to the tables have the symmetry of the corresponding group.

	A	B	C	D	E
A	B	I	E	C	D
B	I	A	E	D	C
C	D	E	I	A	B
D	E	C	B	I	A
E	C	D	A	B	I

FIGURE 6.3 Multiplication table and figure for the symmetric group of degree three

elements because you can adjust a continuously symmetric object by an arbitrary amount and it will still look the same (think of rotating a cylinder). But even for these groups there are clever classification schemes. The idea is to look at little nudges in each of the different directions that leave the appearance of the symmetric object unchanged. All translations along a straight path, for example, can be made up of identical little steps in the direction of the path. All rotations in space can be made up of little rotations about each of three perpendicular axes. The little nudges about each axis can be regarded as the result of an operator, called a *generator*, that acts on the rotation angles. So we can say that the rotations in 3-space have three generators. Classifying all the continuous groups starts with counting the number of their generators, and analyzing the table of multiplications among them. The generators are said to form an *algebra* associated with the group. There is a beautiful connection between these generators and the dynamical quantities that are conserved when a system is invariant (symmetric) under the nudges they effect (described at the end of the next section).

6.10 SYMMETRIES (3): GROUP REPRESENTATIONS

Now you know about groups of operations. So what? Quantum theory provides us with a Schrödinger probability amplitude-plus-phase object – a spreadsheet-like array – whose multiple components lead to multiplicities in the spectrum of objects represented by the field. How can we link such objects to symmetries? The answer lies in an astonishing connection between two apparently distinct topics in mathematics.

Recall that a "multiplication" rule can be defined for the tic-tac-toe arrays called matrices (Chapter 4). It turns out you can always find a *matrix* that stands for or corresponds to each *operation* of a group so the multiplication table for the matrices is the same as the multiplication table for the corresponding group operations. The matrices are said to form a *representation* of the group. It is through the matrix representatives of their groups that symmetries are linked to multiplicities.

All the matrices needed to represent the different operations of a group have to have the same number of rows and columns (the same *dimension*), so "multiplication" among them can be defined. But you could have two different sets of matrices, each set with its own dimension, and the matrices of each set could obey one and the same multiplication table. That is, more than one matrix representation can be found for a given group. Interestingly, *you cannot just choose any dimension you like to represent a given group.* That is, you might have a group multiplication table that can be satisfied by a certain set of four-by-four matrices, but not by *any* set of two-by-two matrices. This explains why Dirac failed to find a relativistic version of the Schrödinger equation for a two-component electron. The symmetry group of Minkowski's space-time (translations and rotations in three space and one time dimension) does not permit a two-dimensional representation for massive spinning electrons. The smallest dimension that will work is four. Dirac did not know about group theory at the time, but discovered that a four-dimensional representation would work by "playing about with the equations." Hermann Weyl and Eugene Wigner shortly afterward provided the full picture in influential books that drew the attention of physicists to the importance of group theory in their new microscopic picture of the world.[17]

Matrix representations of course exist for the generators of the algebras (the "nudges") associated with the continuous groups described in the previous section. The matrix representatives of these *infinitesimal generators* play the role in Heisenberg's original "matrix mechanics" of the conserved dynamical variables. Thus the matrix corresponding to the "nudge" that rotates the system about an axis behaves in matrix mechanics like the component of angular momentum about that axis. In matrix language these variables obey equations that look very much like the classical equations of motion derived from Newton's Law. Recall that it was this form of the quantum equations of motion that Schrödinger and others proved was equivalent to Schrödinger's wave equation.

6.11 THE GAME OF PARTICLE DISCOVERY

The game of particle discovery, therefore, is something like this: Look for multiplicities in the observed particle spectrum. This suggests the number of components of the underlying Schrödinger object. Assume this number is the dimension of a matrix representation of some unknown symmetry group under which the motion of the particles is invariant. Then look for groups that permit a representation with this particular dimension. See what the dimensions are of other permitted representations of this group. If the symmetry is a "real" feature of Nature, there should be new particles whose Schrödinger field has the multiplicity of these other representations. Build an experiment (conceptually, a beam incident on a target followed by a filter and a detector) to observe these new particles. If you see them, you win points in the game. Sometimes, but not always, you win a Nobel Prize.

Each of the world's powerful accelerators during the discovery period were playing fields for this game. Players included experimenters who built the detectors and compiled the particle spectra, theorists who combed the data looking for multiplicities and worked the mathematical machinery to infer symmetries and predict new particles, and accelerator builders who pushed the limits of technology to make brighter, higher energy beams to drive the spray of particle radiation that fed the detectors. Powerful accelerators were built in the United States at Berkeley (Lawrence Laboratory), Brookhaven, Stanford (SLAC), and Fermilab, and important discoveries were made at each. The Atomic Energy Commission, offspring of the wartime Manhattan Project, funded the US machines. Outside the USA, CERN was the most productive laboratory, but important discoveries were made at the other large facilities cited in Note 9. The keys to new discoveries are the maximum energy a machine can achieve, which sets the maximum mass of excitations ("particles") it creates; the brightness or luminosity of its beam, which determines the number of collisions per second; and the efficiency of the detector, which determines how much useful data you get per collision. Together these determine how long an experiment must run to accumulate enough events to check the probabilistic predictions of the quantum models. The main advantage of the machines over sifting through naturally occurring cosmic ray data is that the machines produce the data faster. Much faster. And the initial conditions for the collisions can be known precisely.

I have, of course, oversimplified things. The multiplicities are not so obvious because the symmetries we have discovered are rarely exact. Symmetries are properties of the equations of motion – the Rules of the Dance – not of the actual positions of the dancers. Nature hides her symmetries beneath a heap of contingent circumstances. Because you have to guess at what particles to include in a multiplet, and because you do not know if you have all the data, the analysis rarely points to a unique symmetry. When I describe the final results in the next chapter, you will see that Nature has contrived a scheme of fundamental objects much more subtle and complex than anyone imagined as the big accelerators began to operate at midcentury.

6.12 UNITARITY AND RENORMALIZATION

The multiplicities, branching ratios, lifetimes, masses, and charges of the observed microscopic objects are experimental clues to the right symmetries and the right equations of motion, but there are some additional technical constraints that help narrow the search. Two of these are so important that you should know about them. The first is simple: the thing that satisfies the equation we are looking for is a component of the wave function, a complex number whose (squared) amplitude is a *probability*. Probabilities cannot be greater than "one" (i.e., 100%) nor less than zero. Therefore any equation of motion – or any formula derived approximately from it – that permits the amplitude to dance out of these bounds is wrong. This condition is called *unitarity.* It is more useful than it sounds because the mathematical contortions required to squeeze meaningful results from the complicated equations of the theory can inadvertently lose track of things like relativity, gauge invariance, and unitarity. In practical work these conditions must be checked and any deviations tracked down and understood.

The second constraint on the theory is more subtle, but of the utmost importance. It comes from the logical problem I described at the beginning of this chapter – the third difficulty exposed by Maxwell's theory. Put aside quantum ideas for a moment, and consider again the issue of mass of the fundamental objects.

Imagine the objects are electrically charged particles. Charged particles are linked to forces. When a particle moves, its surrounding field of force does too, evolving according to Maxwell's equations. The energy of the moving field is equivalent to mass. So the moving particle seems to have *more* mass than just the number m that goes into the $F = ma$ equation before solving the equations. If we were to use the *measured* mass m' in $F = m'a$, and solve it together with Maxwell's equations, whose fields contribute to the "F" in $F = m'a$, then we would get the wrong answer. We need to put something *less* than the measured m' into the original equations so when it combines with the effective mass that comes from the moving fields the combination gives the correct measured result. The mass m that actually goes into the equations is called the *bare mass.* Since this differs from the measured mass m', the *dressed* mass, how do we know what it is?

Since the equations of motion are mathematical, we can imagine (with luck and cleverness, and if we have the right equations) that we can *calculate* how the dressed masses depend upon the bare quantities. The result is a graph, a pairing of m' and m. We can read this backward to find m given m', which is what we measure, and that is how we know what bare masses to put into Newton's Law to start with. In the quantum equations, the set of coupled Dirac equations plus field equations, a similar dressing procedure is necessary for the electric and other charges. And the probability field too has to be dressed to preserve unitarity in the face of all this self-interaction. As long as we have only a finite number of such parameters to dress, the process is workable, at least in theory.

This program for fixing the masses, charges, and other quantities in the theory to account for the effect of the dynamic force fields is called *renormalization*.[18] It is challenging enough to think through logically, but much more difficult to carry out in practice. In the Standard Model, both the bare masses and the corrections required to dress them turn out to be *infinitely large* even in the simplest case of electrons interacting electromagnetically. The signs are such that when the two are added the infinite parts cancel leaving a finite result equal to the observed mass. This mathematically horrible situation derives from the possibility that two charged objects may draw arbitrarily near each other, causing the electric force between them to grow indefinitely large according to the inverse square law. These infinities are a *fourth difficulty* embedded within Maxwell's theory that also infects all the other forces, and it is the only one of the four difficulties not completely solved in the Standard Model. The renormalization program is designed cleverly to sidestep the infinities, but the theory still bears their traces in its mathematical bones, if not in the final numbers it produces for energy spectra and detection probabilities. Only an entirely new approach to physical theory is likely to remove the infinities, and indeed the so-called "string theories" promise to do this. Unfortunately, these theories are still visions upon the horizon of the great sea of ignorance, and not yet firmly linked to the connected islands of the Standard Model, which today accounts for nearly all the known fundamental phenomena of the physical world.[19]

Here is the point: Surprisingly, the renormalization program *cannot be made to work for most equations of motion*. It does not work, for example, for Fermi's theory of beta decay described above. It *does* work when the force equations are introduced via the Weyl/Yang–Mills approach. Figuring this out was worth a Nobel Prize in 1999 for Martinus Veltman and his student Gerardus t'Hooft, and it was one of the final results that convinced physicists the Yang–Mills theory might correctly describe Nature after all.

The ultimate goal is to find the set of Nature's microscopic symmetries consistent with these technical constraints. If you know the symmetries then you can write down the correct "single-particle" coupled Schrödinger–Dirac equations for the multi-component object that describes the particle spectrum. Then you can apply the Weyl/Yang–Mills gauge recipe to discover the formulas governing the forces among the various parts of this object. Finally you replace the wave functions in the single-particle equations with field creation and annihilation operators to obtain the starting point for calculations that you hope accurately predict the frequencies of registrations in detectors corresponding to real experiments.

6.13 SPONTANEOUS SYMMETRY BREAKING

Notice carefully that the underlying symmetries of Nature control the system of *equations of motion*, the Rules of the Dance, not the "wave" functions that satisfy them. The wave functions depend on the initial conditions as well as on the equations of motion, and the initial conditions reflect the myriad contingencies of cosmic history. They are most decidedly unsymmetric. We cannot work backward from observations to the underlying symmetry unless we understand how that symmetry is "broken" in the world we observe.

If each of Nature's forces were weak and independent, the spectral multiplets that reveal her symmetries would lie close together in energy, perhaps so close as to appear to be a single line. Atomic spectroscopists often apply magnetic fields whose well-defined direction breaks the rotational symmetry of isolated atoms and splits some of the multiplets so they can be observed. To use the obsolete but visualizable language of Bohr orbits, the energy of the atomic electrons is then no longer the same for different orientations of their orbits with respect to the applied field. The discrepancies appear as separations of the lines within a multiplet. In the simplest case the energy splitting in the spectrum is proportional to the strength of the applied magnetic field that pulls on the little magnets associated with the circulating electrons.[20]

Sometimes, however, the system can break its own symmetry *spontaneously*. The standard example is an assembly of many atoms in a piece of iron, each with its own magnetic moment. At high temperatures, thermal motions continually disorient the little magnets so their fields do not add up coherently and the overall magnetic field within the iron averages to zero. As the temperature is lowered this thermal disorienting effect becomes weaker, and the mutual interactions take over and cause the atomic moments to line up all together to create an overall large magnetic field. This produces what we call a *magnet*, a decidedly macroscopic phenomenon. The spontaneously appearing field randomly emerges along a particular direction in space, breaking the rotational symmetry for each of the atoms within the magnet and splitting their spectral lines into multiplets as described above. Of course no one can predict in which direction the spontaneous field will point. That is a matter of chance, as in the child's game of "spin the bottle." In the late 1950s physicists began to realize that a similar phenomenon might occur in the system of fundamental building blocks of matter.[21]

Atoms and nuclei are made of stable pieces that have intrinsic mass, so a certain amount of energy has to be delivered in a small volume to produce them. They do not pop into existence out of empty space. But we know the electromagnetic field has *no* mass, so it is easy to excite it with arbitrarily small energies. Suppose Nature possessed a massless field similar to electromagnetism, but which – unlike photons – also had a new kind of charge, somewhat like gravity, so that nearby excitations of the field would attract each other. What would happen? Attracting bodies lower their energy by clumping together, so it is energetically favorable for a self-attracting massless field to materialize spontaneously *throughout space* and clump to the extent the ever-present quantum pressure will allow. This new field would in effect *condense out of the vacuum* into a substratum not unlike the discredited ether. Space would no longer be empty, and would no longer possess the symmetry of the bare framework of space plus time plus other internal dimensions like Schrödinger's phase or isospin space. In particular, it would not have the full symmetry of the equations of motion governing the fields. The properties of any other entities that could interact with the condensed field would be altered by its presence, just as the spectra of iron atoms are distorted by the spontaneously appearing magnetic field in a magnet at low temperatures.

I want to make sure there is no confusion between this idea of a *vacuum condensate* with the "zero-point vacuum fluctuations" required by Heisenberg's uncertainty relations and described in Chapter 3. Dynamical fields like electromagnetic radiation always possess components between which the energy oscillates, passing from "potential" to "kinetic" and back again during propagation. These components are complementary in the sense defined in Chapter 4, and cannot vanish simultaneously. Therefore *all* dynamical fields exhibit "zero-point fluctuations," even when their average values are zero when measured over short times and distances. A field that can interact with itself, however, either by a charge it carries or through the intermediation of another field, can precipitate into a ground state in which its average value does not vanish.

It is an old idea that apparently empty space might contain "something" with which we interact so subtly it escapes our notice. Quantum theory now provides us with machinery by which several different kinds of stuff might exist as vacuum condensates. The Standard Model harbors two examples of this machinery, and there is tantalizing evidence for more. It appears that we live as quantum excitations within a vast system of condensed matter whose existence we can infer only indirectly, perhaps by melting or vaporizing small regions of the condensates within the minuscule volumes occupied by high-energy collisions in our great particle accelerators, and observing the effect upon the properties of escaping radiation. Or we might use collisions to excite directly a kind of quivering motion locally in a ubiquitous condensate that would then decay and emit characteristic radiation. Or we might search with telescopes among the detritus of the Big Bang for some evidence of the influence of condensates on the shape and evolution of the cosmos. These are the grand ventures of contemporary science in an era whose technology provides access to previously hidden realms of Nature.

Now it is time to describe that part of the puzzle of matter called the "Standard Model" that physicists have pieced together during the past century. It is not complete, and it is not as pretty as we hoped, but it hangs together and it works. I will explain further technical details as needed, but the language I introduced in the preceding chapter, and the machinery I described in this one should suffice to grasp the key ideas of the Standard Model framework.

NOTES

1. *Three rapid-fire developments.* The discovery of the electron, X-rays, and radioactivity were products of widespread experimentation with gas discharge tubes during the second half of the nineteenth century. Roentgen's apparatus consisted of an evacuated glass vessel with opposing metal electrodes attached to a voltage source. See Weinberg (2003) for a useful history, and Shamos (1959) for excerpts from original papers.

2. *The concept of a "particle".* "... classical Atomism seems to involve four central claims ... 1. A commitment to *indivisibles*, ... 2. Belief in the existence of *vacuum* or 'Non-Being', purely empty space in which atoms are free to move. 3. *Reductionism*: explanation of coming-to-be, ceasing-to-be and qualitative alteration of sensible bodies in terms of the local motion of atoms ... 4. *Mechanism.* ... [which] claims that no body is ever moved except by an external impulse from another, moving body" (Pyle, 1995, Introduction).

3. *The third major awkwardness of Maxwell's theory.* One attractive possibility was that *all* mass comes from the field energy. Kragh (1999) describes the early twentieth-century effort to identify all mass with electromagnetism as "a revolution that failed." Early ideas were plagued with questions about the role of the *ether* supposed to be the medium in which electromagnetism waves. We owe the first clear distinction between mechanical and electromagnetic mass to H. A. Lorentz (1853–1928), who practically invented classical electron theory. Later work was preoccupied with infinities in the theory caused by the inverse relation between the

energy in the field and the size of a charge that produces it – the energy becoming infinite for a point charge. The process of renormalization side-steps this problem, if it can be made to work at all. H. A. Kramers (1894–1952) among the early pioneers of quantum theory was obsessed with this issue, and his work bridged the classical and quantum efforts. See Dresden (1987) for an excellent biography of Kramers. Dirac too was deeply concerned about this problem. It is most troublesome in the theory of strong interactions, where quarks and gluons are very strongly coupled. Most of the mass of nuclei, and therefore of all normal matter, comes from the strong force field (the *gluon* field), and nearly all the rest appears to come from coupling to the Higgs field. See Chapter 7.

4. *The scheme invented by Dirac.* Dirac's original paper, reprinted in Schwinger (1958), is analyzed in Schweber (1994).

5. *Spectroscopies.* Similarities in the energy spectrum of radiation emitted from different objects at different energies allow conceptual tools developed in one domain to elucidate phenomena in others. Victor Weisskopf gave a good nontechnical account of the relationship of atomic, nuclear, and particle spectra in Weisskopf (1968) but the principle extends to even lower energies. Many of us working on laser phenomena during the 1960s studied the older microwave literature, and even nineteenth-century work on acoustics because so many of the interesting optical effects had analogues in these lower-frequency, lower-energy systems.

6. *Constructing equations for new objects.* Pais recounts this conversation he witnessed at a conference in 1961 between Feynman and Dirac (Pais, 1986):

F.   I am Feynman.
D.   I am Dirac. (Silence.)
F.   It must be wonderful to be the discoverer of that equation.
D.   That was a long time ago. (Pause.)
D.   What are you working on?
F.   Mesons.
D.   Are you trying to discover an equation for them?
F.   It is very hard.
D.   One must try.

7. *Cosmic rays.* The Earth is bombarded from all directions by high-energy nuclei, mostly protons, which probably originate within our galaxy. Their

paths are bent every which way by interstellar magnetic fields, so it is not easy to trace their origins. When they approach the Earth their collisions with nuclei of atoms in the atmosphere create showers of other particles, some of which in turn create other showers, etc. The remnants of these collisions can be detected at the surface as *cosmic rays*. Pais (1986) gives a useful account and references. Much lower in energy is a different population of protons and electrons, called the *solar wind*, that boils continually out of the Sun. These cause the auroral displays near Earth's magnetic poles, but do not carry enough energy to create new particles.

8. *Big Science*. The first "big science" projects were astronomical observatories, a famous early example being Tycho Brahe's *Uraniborg*, built by the Danish government in the late sixteenth century on the island of Hven at a cost of 1% of the national budget. This observatory produced the data from which Kepler deduced his laws of planetary motion. Big Science in physics had its origins with Ernest O. Lawrence's sequence of ever larger cyclotrons, constructed at the University of California at Berkeley with private funds until World War II, when it became clear that cyclotron technology could be used to separate uranium isotopes. The war effort launched MIT's Radiation Laboratory, with roots in the privately funded laboratory of Lawrence's patron, Alfred Loomis, and the Rad Lab in turn provided many of the personnel for the huge US Manhattan Project to create nuclear weapons. These large wartime programs were followed by newer high-energy accelerators and the enormous US space program of the 1960s. Oak Ridge National Laboratory director Alvin Weinberg wrote eloquently about the impact of these huge enterprises in essays collected in Weinberg (1967). See particularly Weinberg (1961), where he coined the term *Big Science* in its current sense. See also Kragh (1999) for a good brief history of Big Science issues. I. I. Rabi, one of the founders of the big laboratories, told Pais "You people [the younger generation of physicists] were fortunate to be the grandchildren of the atomic bomb and the children of Sputnik" (Pais, 1986).

9. *Particle accelerators*. The first accelerator to boost particles to energies greater than the mass-energy of a proton (about a billion electron volts or 1 GeV) was the "Cosmotron" at the U.S. Atomic Energy Commission's Brookhaven National Laboratory, so-named because it would produce phenomena previously observable only in cosmic rays. This was followed by the Bevatron (6+ GeV) at BNL's older sister lab at Berkeley, where in 1930 E. O. Lawrence had invented the cyclotron, the progenitor of all circular

accelerators. Stimulated by developments at Brookhaven, European nations cooperated to establish a European Council for Nuclear Research (Conseil Européen pour la Recherche Nucléaire: CERN) near Geneva, which is currently (2010) operating the world's highest-energy machine (the Large Hadron Collider with two opposing beams at 5000 GeV). Soviet bloc nations formed a counterpart to CERN in the Joint Institute for Nuclear Research at Dubna, whose 10 GeV synchrotron was for a while the world's largest. Later the Soviet government built another laboratory near Serpukhov, in Novosibirsk. Both laboratories continue to operate important accelerators. Significant machines exist today also in Italy, Germany, Japan, and China.

10. *Isospin.* Heisenberg first introduced the two-valued parameter that later became associated with the isospin projection in a new "internal" space (1932), but his reasons for doing so were based on a physical picture that quickly became obsolete. Later (1936) B. Cassen and E. U. Condon developed the modern formalism in analogy with electron spin. E. P. Wigner coined the term *isospin* the following year. Pais (1986) is an authoritative source for this history.

11. *Isospin in nuclei.* The nuclei of boron, carbon, and nitrogen, for example, each have an isotope with a total of 12 protons and neutrons: $^{12}B = (7n, 5p)$, $^{12}C = (6n, 6p)$, $^{12}N = (5n, 7p)$. The isospins can be arranged to form an isospin triplet of nuclei with very similar spectra: the two residual neutrons in ^{12}B point down, the two residual protons in ^{12}N point up, and the corresponding n and p in ^{12}C point "sideways." But the n and p can also be arranged to cancel the isospin totally in ^{12}C giving a different state of the same carbon nucleus with a spectrum (a *singlet* spectrum) quite different from the others. These spectral features are observed in experiments. (The two arrangements lead to wave functions for the neutron–proton pair that are superpositions of the individual wave functions as follows: let $n(1)$ stand for the wave function of a neutron at position 1, and similarly for the proton wave function. Then the combinations can be written symbolically as $(n(1)p(2) + p(1)n(2))/\sqrt{2}$ for the "sideways" component of the triplet, and $(n(1)p(2) - p(1)n(2))/\sqrt{2}$ for the singlet.)

12. *"Who ordered that?"* I. I. Rabi (Nobel Prize in 1944 for magnetic resonance), who played an important role in the creation of the large laboratories at Brookhaven and CERN, had an excellent sense of humor and a strong personality that is well documented by his biographers.

I cannot find the origin of this famous question about the muon, but it is repeated by many authorities.

13. *Muons independent of strong nuclear force?* Measurements of the muon magnetic moment (the so-called *g*-2 experiment) at Brookhaven National Laboratory at the turn of the twentieth century showed deviations from the "naively expected" moment that are mostly, but apparently not entirely, accounted for by interactions with other objects in the Standard Model, including the strongly interacting quarks and gluons. The remaining deviations are clues to the existence of matter beyond the Standard Model.

14. *Pauli and Yang.* As Pais tells the story, Pauli had invited Pais to dinner after meeting him for the first time in 1946. "He [Pauli] began to talk of his difficulties in finding a physics problem to work on next," writes Pais, and Pauli said "Perhaps that is because I know too much" (Pais, 1986, p. 314). Pauli was 46 at the time, and he sketched out an isospin gauge theory seven years later, a year before Yang and Mills, but communicated his results only to Pais in private correspondence. When Yang described his work at a 1954 seminar at the Princeton Institute for Advanced Study, Pauli was in the audience. Yang records this exchange in his autobiography: "Soon after my seminar began, when I had written on the blackboard [the starting equation containing the new gauge-field B] Pauli asked 'What is the mass of this field B?' I said we did not know. Then I resumed my presentation, but soon Pauli asked the same question again. I said something to the effect that that was a very complicated problem, we had worked on it and had come to no definite conclusions. I still remember his repartee: 'That is not sufficient excuse.' I was so taken aback that I decided, after a few moments' hesitation, to sit down. There was general embarrassment. Finally Oppenheimer said 'We should let Frank proceed.' I then resumed, and Pauli did not ask any more questions..." (Yang, 1983). Yang was 32 at the time, two years before his work with T. D. Lee that led to a Nobel Prize for parity nonconservation. See below.

15. *Nonlinear optics.* Maxwell's equations for fields in materials like glass contain terms that describe the response of matter to the electric and magnetic forces. For small forces the response is proportional to the force, so a graph of response versus force is a straight line. The stronger fields of laser light can induce a nonlinear response. The resulting *nonlinear optical effects* permit the manipulation of light in much the same way that electronics components manipulate electrons. Equations describing

optical beams resemble Schrödinger's equation, and field theoretical techniques are useful for analyzing them. The gauge machinery was useful but not essential for the problem mentioned in the text. My collaborator, Juan Lam, had done much of the work using more conventional methods. See Marburger and Lam (1979). The point is that gauge invariance is a useful mathematical concept not unique to particle physics.

16. *Symmetries and conservation laws.* The precise relation between symmetries in the energy function and conservation laws was established in great generality by the German mathematician Emmy Noether (1888–1935) in 1918. For technical reasons a mathematical object derived from the energy, called the *Lagrangian*, is a more convenient starting point in the search for symmetries. In the simplest case, the Lagrangian is the *difference* between kinetic and potential energy. In practice, theorists first choose a symmetry and construct a Lagrangian that possesses it, from which the analogue of the Schrödinger/Dirac equations may be derived.

17. *Relation of groups to quantum theory.* In his biography of Wigner, Pais recounts how Wigner consulted John von Neumann in 1926–7 regarding the mathematical treatment of multi-particle systems in quantum mechanics. Von Neumann pointed him toward group theory, which Wigner quickly mastered and applied to other quantum problems. "Thus," writes Pais, "did group theory enter quantum mechanics" (Pais, 2000). Wigner later published a book on the subject in 1931 (Wigner, 1959), which, however, appeared three years after Weyl had published (Weyl, 1930). Pais writes "That sequence of events angered Eugene throughout his life," quoting from a 1963 statement by Wigner that "Weyl had before [my] article no knowledge of it." "It" here can refer only to the multi-particle application because Weyl was already a master of group theory at the time. He did credit Wigner for this application in his book.

18. *Renormalization.* Fundamental physics re-established itself in post-war America with new experimental tools based on the wartime developments in radar and nuclear science, and with new talent that had flowed in from Europe during the rise of the Axis powers. The first great breakthrough, stimulated by a historic meeting of scientists in June 1947 on Shelter Island, New York, was the discovery of practical and reliable mathematical methods to accomplish renormalization. R. Feynman, S. Tomonaga, and J. Schwinger shared the 1965 Nobel Prize in physics for this work. Feynman and Schwinger produced approaches that at first looked as different from

each other as Heisenberg's version of quantum mechanics from Schrödinger's. Freeman Dyson demonstrated their equivalence in a contribution no less important than those of the three Laureates. An authoritative source for these developments is Schweber (1994).

19. *String theory.* At this date, remains more of a program than the kind of definite mathematical structure we know as the Standard Model, but it has great appeal for theorists. It accounts automatically for the gravitational force and accommodates "supersymmetry" between bosons and fermions. Brian Greene does a better job than most at explaining its appeal (Greene, 1999). Lee Smolin and Peter Woit approach string theory with much insight and considerably less enthusiasm (Smolin, 2001, 2006; Woit, 2006).

20. *Splitting of spectral lines in a magnetic field.* Already in the 1860s Michael Faraday had suspected that a magnet might alter the emission spectra of atoms. Pieter Zeeman observed an effect in 1896 that H. A. Lorentz attributed to moving electrons within the atom. By the end of the century, however, higher-resolution spectra showed more complexity than Lorentz's theory had predicted. It was Pauli's struggle to understand this so-called "anomalous" Zeeman effect that led him to postulate the existence of an additional degree of freedom for the electron (later identified as "spin"). I was fascinated by this history in high school, and tried to measure the Zeeman effect as part of a science fair project. The effect was much too small to see with the crude spectrometer I made.

21. *Spontaneous symmetry breaking.* This concept was familiar in solid state phenomena, but its application to particle physics met with considerable resistance, which makes for interesting history. Crease and Mann devote their chapter 13 to the subject, based on interviews they conducted with many of the actors in the 1980s (Crease and Mann, 1996). Spontaneous symmetry breaking in particle physics is today most often associated with the unification of electromagnetic and weak forces, for which Abdus Salam, Sheldon Glashow, and Steven Weinberg received the Nobel Prize in 1979, but the basic idea is much more general, and its application to particle physics began in 1960 with Y. Nambu (Nobel Prize, 2008). Beyond the physics community it is not widely appreciated that what we call "empty space" is filled with condensed matter. A useful reference on the "vacuum," somewhat more technical than this book, is the collection of papers in Saunders and Brown (1991), particularly the essay by I. J. R. Aitchison.

7 The Standard Model

By 1926, when quantum mechanics finally emerged as a coherent theory, the evidence for an atomic structure in Nature was overwhelming. Today, scanning microscopes allow us to "see" chemical atoms one-by-one.[1] They are real. Since quantum theory makes no reference at all to particles, why then do we have all these particulate atoms, ions, and nuclei? How do we account for the success of those stick-and-ball models of molecules, or crystals, or Crick and Watson's double-helixed DNA? The answer is that atomic nuclei are made of stuff that interacts with a force law that gives its assemblages an intrinsic tiny size. The parts are sizeless quantum entities, but the things they make can behave – most of the time – like microscopic particles. The force law and its properties are consequences of the fundamental symmetries of the *Standard Model* which I will now endeavor to describe.

Apart from their tendency to clump in tiny and well-defined sizes, two obvious features mark the list of all the world's building blocks. First, the great disparity in mass between atomic nuclei and electrons, a ratio of nearly 2000 even for hydrogen, the lightest nucleus of all. Thus we have light objects (*leptons*) and heavy objects (*hadrons*). Second, the vast difference in strength between the electric force that holds electrons to nuclei to make chemical atoms, and the much stronger force that holds the nucleus itself together against the mutual repulsion of its electrically charged components. This difference shows up in the energies of chemical versus nuclear reactions – tons of chemicals to measure the energy contained in grams of fissionable nuclei (a gram is roughly the mass of a standard paper clip).[2]

Remarkably, the formal structure of the Standard Model ignores these huge differences of mass and strength. These striking features emerge more or less as side-effects of the particular rules governing the forces among the basic objects. The rules, moreover, all spring from the same underlying requirement that Nature's laws of motion must not depend upon arbitrary choices of reference frames for internal "spaces." This is a symmetry requirement: the laws must "look the same from different points of view." And it is the symmetry of the internal spaces that determines the pattern of multiplicities – the structure of the Standard Model. Once the intrinsic symmetry is established, (almost) everything else follows. As C. N. Yang insisted, *"symmetry dictates interaction."*[3]

All the entities of the Standard Model respond to the slight but ubiquitous force of gravity whose source is energy itself. They differ, however, in their response to at least three other forces: *electromagnetic* and *weak forces* (they link to form the *electro-weak force*) and a *strong force*, each generated by a different kind of charge. These forces are like Maxwell's electromagnetism – dynamical fields that can carry their own separate energy and momentum and come in Planck quanta like the photons of electromagnetic radiation. Quanta of force fields are known by the generic name of *bosons*, in accordance with the symmetry of their wave functions (Chapter 5). The leptons and the fundamental components of hadrons, called *quarks*, are *fermions* with antisymmetric wave functions. Gravity, the weakest but most familiar force of all, remains outside the Standard Model. Ingenious ideas have been proposed to incorporate gravity in the overall quantum framework, but the theories are excruciatingly incomplete. The Standard Model includes a fifth kind of force, not yet observed directly in experiments but essential for the entire framework. This is associated with a *Higgs field*, which in our era of cosmic evolution is thought to have precipitated from the vacuum through a poorly understood self-interaction to form a pervasive condensate throughout the universe that influences all other matter. The Standard Model seems to be a fragment of a larger framework. Tantalizing evidence from astronomy points to other objects and forces, but their origin remains obscure.[4]

Particle names and flavors. What physicists call "elementary particles" generally appear as narrow features – "lines" – in the spectrum of mass/energy spewed out from collisions in accelerator experiments or cosmic rays. So naming the particles should be as simple as naming these features. Unfortunately *each "elementary particle" we actually observe is a combination of all the pieces of the Standard Model* (Chapter 6). The fundamental entities in the theory interact with each other to produce the observable features and are therefore not uniquely related to the peaks of the energy spectrum. In most cases, however, they align closely enough that physicists use the same names for both. The fundamental fermions are said to be in definite states of *flavor* to distinguish them from the somewhat different definite states of *energy* that we observe. As a result, particle names do double duty. They label observable excitations of stuff that have definite energies, and they also label the flavors of building blocks that appear in the scheme of the Standard Model. Physicists tell which is which by context or by using different symbols in the mathematics.

The metaphor of unique particles that can bear different attributes like charge and mass is psychologically powerful because it fits our everyday picture of large objects. Thus amber beads have shape, hue, and mass. But this metaphor misleads at the smallest scale. Quantum physics deals with wave functions that refer to specific detectors tuned for various attributes. A wave function for an object with definite mass/energy may be a superposition of wave functions each with a definite but different flavor, and *vice versa*. The attributes "energy" and "flavor" are incompatible with each other and are not exhibited simultaneously with well-defined values by the underlying stuff. They correspond to different frames of reference in Hilbert space. We are free to choose a standard set of underlying pieces from which all the others may be constructed by superposition just as we can represent waves in space either by their values at each location or by the values of the amplitudes and phases of their component sinusoidal waves (Chapter 3). The pieces that emerge naturally from the theory possess well-defined *weak charge*, and flavor names denote weak charge states. This lack of definiteness about names is not a minor technical detail, but a chief characteristic of the quantum framework.

7.1 LEPTONS

The electron/positron is not the only lepton. Three families of leptons have been found, each with two members or partners, one electrically neutral the other with negative charge (Figure 7.1). All have one-half unit of intrinsic spin – the smallest possible nonzero value – and the partners have equal and opposite values of the weak charge, known for historical reasons by the thoroughly confusing name *weak isospin*. The intrinsic spin and the weak charge are correlated in an amazing way that I mentioned at the end of Chapter 4, and will describe later. Because the three families closely resemble each other, and each family is heavier than the next, physicists arrange them in *generations*. The leptons can carry weak and electromagnetic charges, but not the charge for the strong force.

The conventional names of the leptons (and the lepton flavors) in the three generations are the *electron* and the *electron neutrino*, the *muon* and *muon neutrino*, and the *tauon* and *tauon neutrino*.[5] The masses of the neutrinos are quite small relative to the electron (hundreds of thousands of times smaller), hence the diminutive "-ino" reflecting the linguistic heritage of Enrico Fermi who named it. Just now (2010) we do not know the absolute neutrino masses very well, but we do know they are all different and that they lie in the range of one ten-millionth to 5.6 ten-millionths of an electron mass.[6] The

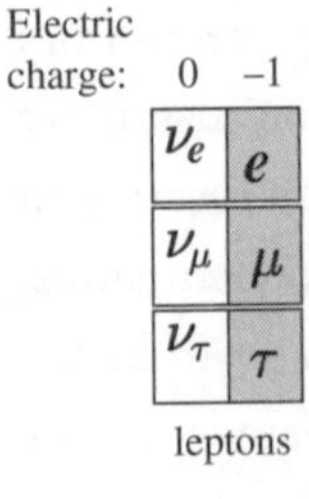

Electric charge: 0   –1

ν_e	e
ν_μ	μ
ν_τ	τ

leptons

FIGURE 7.1 *The leptons* occur pairwise in three families shown here with the mass increasing from top to bottom, as in the periodic table of chemical elements. Neutrinos are designated by the Greek letter *nu* (ν). Objects in shaded boxes are electrically charged. All objects in the same column have the same charge. In the diagram, the pairs are displaced vertically to indicate a difference in weak charge (see text).

neutrinos are electrically neutral but carry the weak charge. I wish I could say they are essentially the same as the charged electron-like leptons, but they might be subtly different. Since they have no charge, it is difficult to distinguish between neutrinos and anti-neutrinos. It is an open question today whether the neutrinos are their own antiparticles, or whether their antiparticles are distinguishably different. The Standard Model grew up assuming massless neutrinos, but it can easily accommodate neutrino mass in exact analogy with the electron. The other leptons all have the same electric charge as the electron.

Regarding the heavier charged leptons, the muon is 207 times and the tauon 3490 times more massive than the electron, which is itself a petite 9.1×10^{-31} kg corresponding to a mass energy of 511 keV. The names come from the Greek letters *mu* (μ) and *tau* (τ) picked more or less arbitrarily to designate these objects in formulas. As far as we know they are all "fundamental" (not made up of other pieces), but the muon and tauon are unstable. The muon, when it is standing still, decays in about 2 µs to an electron and two neutrinos. When moving its lifetime is stretched, depending on its speed, by relativistic time dilation. The very heavy tauon decays a million times faster through several different branches, all eventually ending up as one or two electrons and assorted neutrinos.[7] The neutrinos themselves do not decay, but they can mix together as described below.

Neutrino oscillations. As explained above, either flavor or mass/ energy can be used to distinguish among the different fundamental objects. Nothing in the theory describing these objects guarantees that measurements of one are compatible with measurements of the other. Well-defined states of mass could imply poorly defined states of flavor, and *vice versa*. Observations show this is indeed the case.

Imagine measuring the masses of neutrinos that are born in a nuclear process that gives them definite flavor, say "electron-type." If definite flavor implies indefinite mass, you will see a spectrum of all three different neutrino masses among the measurements. Because the energy of a system – including its mass energy – determines the frequency with which its wave function oscillates, the wave function for a neutrino with definite flavor will therefore be a superposition of waves oscillating at three different frequencies. As time passes "beats" will occur among the different frequencies. That is, interference among the three differently oscillating components will spoil the original mix that gave just the right flavor. Thus a neutrino wave function that initially had all its energy components phased just right to give electron-type flavor would evolve in time into a superposition of electron-, muon-, and tauon-type flavors. We have seen something like this before: An electron starting out with a definite position has a wave function with a spread of momenta that leads as time passes to a spread in subsequent measurements of position (the sharply peaked wave function *disperses* as it travels). A similar phenomenon occurs when pendulums with slightly different lengths all swing from the same support (e.g., a horizontal bar). You can start them all swinging together, but as time passes they begin to move "out of phase." The electron–neutrino wave function has three different mass components, and subsequent measurements will reveal the effect of three closely spaced frequencies beating together. This is the phenomenon of *neutrino oscillations*.[8] Similar oscillations can occur among other Standard Model pieces.

Neutrino oscillations solve a puzzle about the Sun. The Sun's energy comes from a chain of nuclear reactions that generate electron-flavored neutrinos in definite amounts that can be calculated. Since they have no electrical or strong nuclear charge, neutrinos scarcely interact at all with other matter (they interact via the *weak force*, see below). So they should be good probes of the Sun's interior, passing directly out from their origin to our detectors (which are necessarily huge – bigger than barns – to capture such elusive quarry). But for decades, measured neutrino fluxes from the Sun always came up short of theoretical predictions. We now realize the neutrinos were not missing at all, but just converted through the oscillation mechanism into neutrinos of other flavors that our detectors were not tuned to register. Changing the detection method to see other flavors made the "missing" neutrinos visible.

7.2 QUARKS

The heavy hadrons are all made of *quarks*, which, like the charged leptons, are intrinsic spin one-half objects described by four-component wave functions obeying Dirac's equation. They also come in three generations of families with two members each, like the leptons.[9] The three pairs of quarks in order of increasing mass are called (and the corresponding flavors are labeled) *up* and *down*, *charm* and *strange*, and *top* and *bottom* (or *u*, *d*; *c*, *s*; *t*, *b*; with primes to distinguish anti-quarks: *u'*, *d'*, etc.). The quarks carry all three kinds of charges – electric, weak, and strong – and unlike the leptons they can clump together under the strong force to make long-lived objects with definite size and mass, the most stable of which are the nuclei of atoms. The first member of each quark pair, often called the "up"-type partner, carries a positive electric charge two-thirds that of the electron. The other "down"-type partner in each pair has a charge one unit less (like the leptons again), or negative one-third.

The strong force, which acts only among quarks – thus distinguishing them from leptons – is generated by a new kind of charge that differs from electricity in having *six* rather than *two* states. Three of these are the so-called *color charges* found on the quarks (call them *red*, *blue*, and *green*), and the other three are the charges found on the anti-quarks (call them *anti-red*, *anti-blue*, and *anti-green*).[10] Compare with the negative electric charge on the electron, and the positive charge on the anti-electron (positron). Since each quark comes in three "colors" it would be more accurate to count $3 \times 6 = 18$ different kinds of quarks. The colored versions are all so similar it is easier to think of them as just different states of the same underlying object, so we usually speak of only six quarks and their anti-quarks. To be consistent we should abandon the term "positron" in favor of "anti-electron" but in physics as elsewhere tradition often defeats consistency.

Because the color force is so strong, its field – whose quanta are called *gluons* – strongly influences the properties of the hadrons, so it is more accurate to say that hadrons are made of quarks *and* gluons. Like

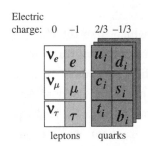

Electric charge: 0 −1   2/3 −1/3

leptons   quarks

FIGURE 7.2 *The quarks*, like the leptons, occur pairwise in three families. Dark shading indicates objects that have color as well as electric charge. Subscripts *i* on quarks indicate *red, green,* or *blue* corresponding to the three layers shown. Quarks possess weak charge in the same pattern as the leptons, as explained in the text.

the leptons, the quarks appear to be fundamental, not made of more elementary parts.

Mesons. Electrically neutral objects can be built up from charged components by combining charges with anti-charges – positive balancing negative as in the neutral hydrogen atom with its proton and electron. "Color-neutral" structures can also be made this way, and the set of hadrons called *mesons* are such colorless combinations of a quark and an anti-quark. Nature, however, is exceptionally color-blind. The meson wave functions are superpositions of *all three* colorless possibilities, so none is favored. For example, the positively charged *pion*, whose symbol is the Greek *pi* or π^+ with a plus sign for the charge (Chapter 6), has a wave function that is a superposition of red, green, and blue *up* quarks, each paired with an anti-red, anti-green, and anti-blue *anti-down* quark. A common shorthand notation for the wave function suppresses the colors: $\pi^+ = ud'$. The letter u stands for the wave function of the up quark, and d' stands for the wave function of the anti-down quark. So the pion wave function π^+ depends on the *two* locations and spins of the constituent quark wave functions, not just one "pion location." A more complete notation showing the colors but not the spins and locations, is $\pi^+ = (u_r d'_{r'} + u_g d'_{g'} + u_b d'_{b'})/\sqrt{3}$. (The factor $1/\sqrt{3}$ is inserted so the wave function π^+ acts like a probability amplitude if the u's and d's do (Chapter 6). Thus *Nature is symmetric in color-space*: You can permute the letters r, g, and b any way you like in the expression for π^+ and the sum remains the same. A measurement of the strong charges of the two quarks that make the particular pion that has positive electric charge might give a red *up* and an anti-red *anti-down*. But it could equally likely give a blue *up* and an anti-blue *anti-down*, or a green *up* and an anti-green *anti-down*. (Such measurements are not practical but we must imagine them so we can apply the principles of quantum theory.)

Many different mesons have been observed, their quark and anti-quark either hugging each other closely in their ground state, or spinning or oscillating with respect to each other just like the two atoms of nitrogen or oxygen in the tumbling diatomic molecules that make Earth's atmosphere. Unlike the chemical molecules, all mesons are short-lived, decaying eventually to electrons or positrons and a variety of neutrinos. The most stable are the charged *pions* $\pi^+ = ud'$ and $\pi^- = du'$ which, if they are standing still, survive for only 26 billionths of a second before decaying to a muon and its corresponding neutrino. All the other mesons, including the neutral pion, decay at least billions of times faster. The neutral pion wave function $\pi^0 = (uu' + dd')/\sqrt{2}$ is an equal mixture of uu' and dd', a combination dictated by its role as the "sideways" component of an isospin triplet (see Note 11 in Chapter 6). Although not a fundamental internal symmetry, the original isospin concept is useful for quarks as well as for nuclei. Today it is called *flavor isospin* to distinguish it from the more fundamental *weak isospin* discussed below. The intrinsic spins on the quarks of these lightest mesons (i.e., the real "spinning spins," not the metaphorical isospins) are "anti-aligned" to cancel, so the total intrinsic spin of the pions is zero. *Aligning* the intrinsic spins on each quark produces another isospin-one triplet of mesons called ρ (Greek *rho*), whose intrinsic spin is also "one" and which is more than five times more massive than the pion. The business of keeping track of the different kinds of quarks, spins, and charges looks more tedious here than it actually is. Physicists learn a few simple rules that let them reconstruct the full formulas for the wave functions as needed, and they rarely write them out completely.

Two more light electrically neutral mesons called η (Greek *eta*) and ω (Greek *omega*) can be constructed from the u and d quarks, namely the intrinsic-spin-zero and "one" states of the combination $(uu' - dd')/\sqrt{2}$. This differs from the combination that gives the sideways component of the pion triplet, and corresponds to an isospin singlet. All the mesons described by wave functions that substitute s for d in these combinations have also been observed, and so have even heavier mesons containing c's or b's. Their properties can be found in Nakamura *et al.* (2010). These data of *hadron spectroscopy* were essential for working out the scheme of the Standard Model, and continue to give important information for what lies beyond.

Baryons. There is another way of achieving color neutrality with strongly charged objects. The combination of red, green, and blue charges produces a color-neutral object just as mixing light of these three colors makes the neutral color white. This is the property of strong charge that motivated the color metaphor in the first place. Thus the color-neutral proton and neutron are each made of *three* quarks. The abbreviated notation for their wave functions is $p = uud$, $n = ddu$, and a notation that makes the color charges explicit is, for the proton, $p = (u_r u_g d_b - u_r u_b d_g + u_b u_r d_g - u_b u_g d_r + u_g u_b d_r - u_g u_r d_b)/\sqrt{6}$. The pattern of plus and minus signs is necessary to preserve the overall antisymmetry of the quark wave function, as explained in Chapter 5. As above, the position and spin variables are suppressed in this notation, so for example $u_g u_b d_r$ really means $u_g(x_1, y_1, z_1, s_1)\ u_b(x_2, y_2, z_2, s_2)$ $d_r(x_3, y_3, z_3, s_3)$. The anti-proton is a combination of the corresponding anti-quarks. Hadrons made like this from three quarks are called *baryons*. It is this composite nature of the hadrons that makes nuclear physics so complicated. The protons and neutrons that appear to be building blocks of nuclei are actually clusters of quarks (or rather quarks and gluons – see below). In the Standard Model, atomic nuclei are surprisingly complex objects whose behavior is difficult to calculate from first principles, and which we do not yet fully understand. How, for example, do the various sources of angular momentum in the proton – the spins and orbital motions of quarks and gluons – add up to the observed value of ½? This is the kind of question for which the large detectors at the world's high-energy accelerators provide data.

Hadron spectra from collision experiments display numerous baryons and their excited states. As for the mesons, different orientations of the intrinsic spins of the three quark components give baryons with different energies/masses. That is, different choices of the spin coordinates s_1, s_2, s_3, in the expression above for the proton wave function will give different baryons, so the abbreviated notation above should really be enlarged to include the spin coordinate. A convenient way to do this is to append a little arrow indicating whether the quark spin is up or down (the only choices for a spin-1/2 object). Then the proton wave function looks like $p = u\uparrow u\uparrow d\downarrow$ with the misaligned spin paired with the down quark. This is the lowest energy baryon, and it is stable, as far as we know. All other baryons decay to the proton except for the neutron $n = d\uparrow d\uparrow u\downarrow$ when it is bound in a nucleus, and then it is stable too as explained below. The baryon with the same quark content as the proton but all the spins aligned, $u\uparrow u\uparrow d\uparrow$, is one of a multiplet known as Δ's (Greek *delta*). It is not a stable particle, but shows up as a strong peak in the energy spectrum from the scattering of pions from protons. There are four of these: $\Delta^{++} = u\uparrow u\uparrow u\uparrow$, $\Delta^{+} = u\uparrow u\uparrow d\uparrow$, $\Delta^{\circ} = u\uparrow d\uparrow d\uparrow$, $\Delta^{-} = d\uparrow d\uparrow d\uparrow$. The additional mass/energy required to align the spins in these baryons is larger than the mass gained from replacing one of the down quarks with an unaligned strange quark, and some of the resulting multiplets of "strange baryons" are less massive than those made entirely from u and d. Not every combination of spins and quark symbols makes a possible baryon because the final result has to preserve rotational invariance in ordinary space, and obey the symmetry rules explained in Chapter 6. Hadrons having more than three quarks are theoretically possible, but evidence to date indicates they do not exist.[11]

Quark masses. Quarks have masses, but measuring them is awkward. For reasons explained below quarks do not appear by themselves, but only in color-neutral combinations like the mesons and baryons. Here the problem of distinguishing between the quark masses and the energy contained in their binding fields becomes acute. The measured proton mass, for example, appears to be *mostly* from the force fields and very little from the quarks themselves. Since we cannot isolate quarks, we have to adopt an indirect recipe to define their masses. One approach is simply to assign a fraction of the total mass of a meson or a baryon to each constituent quark, ignoring the binding field altogether. The result is known as the *constituent mass*. Alternatively, we could appeal to some theory that contains bare quark masses and match to experiments to infer their values. The renormalization process described in Chapter 6 then gives a value for the "dressed" mass the object would have if it could be isolated. I will call this the *inferred* quark mass. (Physicists use terms like *running mass*, *current mass*, and *pole mass*.) However they are defined, the masses of u and d are similar, and those of u, d, and s are much smaller than c, b, and t. In terms of the electron mass, typical values for the inferred masses are:[12]

$$m_u = 5, \, m_d = 10, m_s = 196,$$
$$m_c = 2540, m_b = 8320, \, m_t = 337,000.$$

The three heavy quarks have masses comparable to the nuclei of atoms.

As for the leptons, the quark masses are not uniquely associated with the quark flavors, so we have *quark oscillations* exactly analogous to the neutrino oscillations described above. Unlike neutrinos, quarks cannot be observed directly, and so the oscillations are observed in the decay modes of composite objects such as the mesons. The heavier quarks are all unstable, decaying, if left to themselves, by emission of weak force bosons ultimately to u or d quarks. The weak bosons are themselves unstable, and rapidly decay to other Standard Model pieces.

To summarize, we have six quarks (or 18 if you count the colored ones separately), six leptons, their anti-objects, and the bosons that carry the forces among them, which already seems rather complicated, especially since only two quarks (*up* and *down*) and one lepton (the *electron*) are needed to make all the chemical atoms. All these other pieces are not quite superfluous because they appear to be needed to make the theory logically consistent. Except for one more that I will describe later, these objects, plus rules for their interactions, comprise the *Standard Model*, which correctly simulates nearly all features observed to date in the particle spectrum. They do not, however, account for all the matter that we think exists in the universe. Astronomical observations of galactic motions imply the existence of *dark matter* that presumably interacts with other matter only through the weak force and gravity. The masses of neutrinos, of which there are plenty in the heavens, are too small to account for the observed effects. Other pieces may yet be required to resolve technical puzzles in the mathematics.

Who can resist a smile at this over-abundance of fundamental parts? And we have not yet seen the oddest of this collection of odd things. It cries out for explanation. Newton's admonition to "frame no hypotheses" seems ridiculous in the face of such an extravagant proliferation. This flamboyant "Standard Model" is not yet entirely satisfactory from a mathematical point of view, but its ability to account for nearly every observation is astonishing. It tells us *how* matter assembles itself at the microscopic level, but we cannot help asking w*hy?*

7.3 FORCES

The basic fermions of the Standard Model, 3×6 quarks and 6 leptons, all resemble the electron. They each have two components of spin, and in the absence of interactions their wave functions each obey Dirac's four-component version of Schrödinger's equation which pairs objects and anti-objects in one multi-component expression. They differ only in their masses and in the types of charge they bear: electric, weak, or strong. Perhaps they are not really different, but just different manifestations of the same underlying entity. Against this uniformity among the fermions stand in contrast profound differences in the behavior of the forces binding them. And yet these forces all arise from the same Weyl/Yang–Mills mechanism of gauge-invariance generalized for the particular multi-component fields of the basic fermions.

7.4 ELECTROMAGNETISM AND QED

If it were the only one, the simplest force in Nature would be electro-magnetism. The weak force is alarmingly complicated and mixes intimately with the others to make them complicated too. Its weak-ness, however, means these complexities can often be ignored and particularly in phenomena of human scale. Setting aside weak forces, we can "explain" electromagnetism in a quantum description as resulting from a dial-like phase dimension in the geometry of nature in which only relative phases matter (Chapter 3). To repeat the main point: the origin from which the phase of the wave function is meas-ured may be chosen arbitrarily at different locations in space without affecting the predictions of the theory, as long as an electromagnetic potential exists that can be adjusted by a compensating "gauge trans-formation." This is called *local gauge invariance,* but it might just as well be called the *principle of relativity of quantum phase.*

The source of electromagnetism is called *electric charge,* which is conserved as a consequence of the gauge invariance (Chapter 6). The fundamental field of electromagnetism is the four-vector potential with components A_μ (short for A_x, A_y, A_z, A_t) that connect the change in phase with steps in each of the four directions in space-time for an object with electric charge. The actual phase change equals the electric charge times the potential times the length of the step in the direction of the potential vector. The mathematical machinery of gauge-invariance produces equations for $A\mu$ that imply Maxwell's equations. (Physicists use Greek subscripts to stand for the four com-ponents of space-time, Latin for the three of space.)

The bosons whose registration in detectors signifies the presence of the electromagnetic field, or A, are called *photons.* Photons register one unit of intrinsic spin, twice as much as the quarks and leptons, and no mass. "No mass" means you can make photons with as little energy as you like, but it also means they must move at the speed of light. This in turn implies that only two of the triplet of states ordinarily associ-ated with a spin-one object can be observed. You will have to take my word for it that these are logical consequences of Maxwell's equations and have nothing to do with quantum theory *per se*.

Polarization. Photon spin reveals itself in the property of light called *polarization*. Light from most natural sources is a random mixture of polarization components, which masks their effects. Eyeglasses for 3-D movies have "polarizing" filters that block one of the two components – a different one for each eye so the two eyes see different images. Sunlight reflected from surfaces or from the sky is partially polarized, so correctly oriented polarization filters in sunglasses reduce glare. Polarization and other properties of photons are described by Maxwell's equations, which play a role in theory similar to Schrödinger's or Dirac's equations for the wave functions of the fermions of the Standard Model.[13]

Modern quantum theory has its origins in the study of electromagnetism tied to electrons and positrons, a coherent theory in its own right called *Quantum Electrodynamics* (QED). The fundamental equations of this theory are Dirac's equation linked to Maxwell's equation through Weyl's concept of gauge-invariance and suitably modified for Step 2 of the theory as explained in Chapter 5. QED is the most accurate physical theory known, predictions matching observations to a dozen decimal places. This makes it reasonable to invest in computer simulations of the electromagnetic, mechanical, and chemical properties of matter because we know the QED theory contains within it a reliable account of normal matter at moderate scales and energies. New phenomena can sometimes be anticipated by examining the equations of the theory "without looking" at Nature. Nature being enormously complex, however, simulations of constructs with many atoms are necessarily incomplete, even with the largest supercomputers. The trick is to replace the exact equations with approximations that are simpler, retaining just enough complexity to simulate the essential features.

Digression on reductionism in science. With very few and well-defined exceptions, QED accounts for the microscopic basis of all human-scale phenomena. It is effectively the end of the reductionist chain of explanation for nearly all the material things significant in daily life (not counting the effects of gravity). From the physicist's point of view, no further explanation seems necessary, or even possible. That view, however, is profoundly deficient. Yes, the reductionist chain is complete for all humanly practical purposes. But no, this by no means removes the need for and possibility of further explanations of natural phenomena. The "reduction" has been accomplished only with respect to the fundamental equations of motion – the Rules of the Dance – and the few essential Standard Model components to which the Rules apply. That leaves out the crucially important initial conditions without which the Rules are useless. How the stage is set, the number of the actors and their dispositions, determine the play as much as the rules by which the actors move and interact.

The inadequacy of physical reductionism is most apparent when we contemplate the phenomenon called "life," and life's complex layers from molecules to cells to organisms, consciousness and societies. In physics language, the challenge of life phenomena lies first in discovering initial conditions – the particular kinds and arrangement of atoms (as in the structure of a set of proteins snarled in a cell) – second in understanding the interactions and motions of the components, and third in appreciating the significance of these to the behavior of the whole organism. Sometimes we can isolate a subset of a real system small enough to simulate with the known physical laws and the largest computers. I have often watched the fascinating results of supercomputer simulations unfolding on a computer screen, intricate and surprising, and my colleagues and I would exclaim "What's that? Why is that happening?" We might agree that simulations give a reductionist accounting of phenomena, but except in the simplest cases the computed results do not *explain* them. Compared with the infinite variety of initial conditions of the inconceivably immense numbers of elementary excitations we call particles, the complexity of the Standard Model is child's play. Progress in biology requires considerably more than QED.[14]

QED and "Feynman particles." QED predicts in detail how systems of electrically charged and spinning objects move, and estimates with extraordinary accuracy the excitation energies and spectra of light emitted by chemical atoms, not to mention the properties of materials and chemical reactions. In QED, electrons and photons do not appear as particles. There are only the probability amplitude fields that predict detector statistics. The mathematical structure of QED, however, lends itself to visualization through yet another concept for which physicists use the word "particle." The new concept is related approximately to macroscopic particles, but corresponds exactly to pieces of the QED calculational machinery. These "mathematical particles" trace world-lines in a graphical display of a calculational method invented in the late 1940s by Richard P. Feynman (hence *Feynman diagrams*).[15] In this display, *photon lines* begin and end at knots called *vertices* on continuous (i.e., never ending) electron–positron world-lines. Physicists often speak of these lines as if they were the world-lines of real particles, but the probability amplitude is a sum over an infinity of different diagrams, so a unique physically observable particle cannot be associated with any one of them.

As a guide to intuition about Nature, Feynman diagrams are most useful when the forces among objects are weak. This is the case for the weak and electromagnetic forces, but not in general for the strong color force. In the approximation scheme of QED, "particle" motions are first calculated (i.e., Dirac's equation solved) ignoring the forces altogether ("zeroth" approximation), then those motions are used to calculate a first approximation to the forces, which is in turn

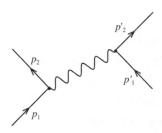

FIGURE 7.3 *Feynman diagram* for a term in the series whose sum gives the probability amplitude to register two interacting electrons with momenta p_2, p_2' if they began with momenta p_1, p_1'. In the Feynman interpretation, the wavy line represents a photon radiated by the first electron and absorbed by the second. The first electron recoils and the second is scattered by the photon. See Note 15.

used to calculate a correction to the motions, and so forth, for ever higher degrees of approximation. You can see that the higher approximations will all be expressed in terms of the zeroth-order expressions for the free "particles" and these correspond to the line segments of the diagrams. Such a process generates an infinite sequence of mathematical expressions or terms whose sum hopefully gives a correct representation of the motions and forces. Each term can be represented by a Feynman diagram, each of whose parts corresponds exactly to an algebraic factor in the term. If the forces are weak the simplest diagrams give an intuitive visual picture of how the various lowest-order parts interact to produce a particular observable phenomenon. Since the correspondence between diagrams and terms is exact, quantitative results can be obtained efficiently by graphical manipulations of the diagrams. Feynman's ability to calculate quickly using his diagrams astonished his colleagues and motivated rapid acceptance of diagrammatic methods in quantum theory.

We have had 150 years to understand electromagnetism, but only a few decades for the strong and weak forces. The visualizable machinery of QED is credible only because the forces carry a small amount of energy compared with the objects they influence. This is mostly not true for the strong force whose geometrical intricacy, moreover, far exceeds that of the interlocking fields of electromagnetism. Consequently the other forces of the Standard Model are not as well explored nor as easily visualized as electromagnetism. You will have to be satisfied with assertions in the following paragraphs supported only by suggestive ideas that almost certainly are not entirely wrong.

7.5 THE STRONG FORCE AND QCD

The strong force, as noted, operates among quarks and originates from three-valued color charge and anti-charge. In Nature we see only neutral combinations of the quarks that bear these charges, just as we normally see only neutral atoms. An atom that loses an electron (in a violent collision, for example) will soon pick up another, restoring overall neutrality. Or it will draw toward itself the electrons attached to nearby atoms, sharing out or *shielding*, as it were, the effect of its charge among the surrounding atoms. Quarks bind mutually in such a way that their color charge is *never* observed to be uncompensated. In Nature we find only "white" hadrons.

For years, physicists tried to derive the strong force using the wrong symmetry. At mid-century, the isospin symmetry looked promising (Chapter 6). It was based on the "two-ness" of the proton–neutron pair. The first quark theory in 1964 expanded the isospin concept to accommodate the "three-ness" of the lightest quarks: *up, down,* and *strange.* Both theories have their value, but neither captures the fundamental symmetry of the quark interactions. The strong force arises from a different "three-ness," that of color charge, and the fact that Nature's laws of motion are symmetric in what we may as well call *color space.* That is, you may change the labels of the color charges any which way at each point in space-time and the equations of motion will still read the same. Or more accurately, you can choose a framework of axes in color space oriented however you like. The quark wave functions have three color components, one along each axis, which will be different depending on how you choose the axes. (Each component is itself a four-component Dirac spinor because we still have to account for the spin and particle/antiparticle attributes.) The Weyl/ Yang–Mills idea then implies that a new kind of "vector potential" – and accompanying force – is needed to compensate for the effects of arbitrary choices for the color axes as you pass from one point of ordinary space-time to another. This is how the Standard Model explains the strong force.

It took a long time for theorists to figure out that the quarks were color triplets, and the right symmetry was the one generated by rotating these triplets in a new set of internal "color dimensions." The relevant symmetry group is called $SU(3)$, with eight generators as opposed to the isospin group $SU(2)$ with three generators. The notation is explained in Note 16 for the dedicated reader. (See also Chapter 6.) Recall that the symmetry that gives rise to electromagnetism in QED, called $U(1)$, has only the single generator that nudges the "dial" of the

Schrödinger phase. In analogy with QED, the theory of objects moving under the influence of color charge is called QCD – *Quantum Chromodynamics*.

Gluons. The bosons that carry the strong force are called *gluons*.[17] Like photons, they have no mass, no electric charge, and one unit of spin. Unlike photons, which carry the electromagnetic force but are not charged themselves, the Weyl/Yang–Mills procedure applied to the richer $SU(3)$ symmetry implies that *the gluons do carry strong color charges*. This has profound consequences. Consider the implications of the fact that color charge, like electric charge, is conserved. If a colored quark radiates a differently colored gluon, it must change its own color charge to compensate the charge carried away by the gluon. It might change from red to green, for example, in which case the gluon must carry a *double* "red + anti-green" charge. That way the anti-green part cancels the quark's new green, and the red part assures that the net charge of the quark + gluon remains red. Thus every gluon has *two* charges in a kind of *color dipole*, one of which is anti-red, anti-blue, or anti-green, and the other is red, blue, or green. There are eight different kinds of gluons that matter (one to "gauge" each generator of the $SU(3)$ color symmetry), each of which has twice the amount of color charge as a quark.[18] They also spin twice as much, which suggests stronger "color magnetism" than the quarks. Therefore we expect the forces among gluons to be stronger than the forces among quarks or between a quark and a gluon.

The gluon condensate. This gluon self-interaction has bizarre consequences. Gluons, like photons, are massless so you can make them with as little energy as you like. Because they attract *themselves*, excitations of the gluon force field can lower their collective energy by clumping. If they clump too densely, quantum pressure will push them apart again. Calculations suggest, however, that there is a clumping size for which the attractive force may lower the energy of a mess of gluons below the energy even of empty space. That is, a volume full of clumped gluon stuff is more likely to occur than one with no gluons at all. Since gluons interact strongly with quarks, the substance of this

QCD vacuum condensate must include some quarks as well, or more precisely, some colorless combination of quarks, anti-quarks, and gluons. We expect the lightest quarks, u, d, and perhaps s, and their antiparticles to dominate in this complicated mixture, because the other more massive quarks would be harder to create without exceeding the energy gained by clumping. Where space contains very hot matter, such as shortly after the Big Bang or in a collision between nuclei in an accelerator, this complicated condensate would be heated and the attractive color force that causes it would not be able to overcome the resulting thermal pressure. But as things cooled after the Big Bang, the gluon–quark mix must have condensed into something like a sea criss-crossed with an intricate pattern near the critical clumping size. The clumping occurs in time as well as in space, which means lumps continually form and dissipate. What we call "vacuum" is really filled with a restless sea of coalescing quarks and gluons! Call it the *QCD vacuum*.[19] This sea strongly affects the quark clusters that assemble into protons and neutrons which comprise most of normal matter. Just how this works is still a subject of investigation. Theorists have conceived several approximate pictures of the QCD vacuum, none entirely satisfactory.

In one picture the QCD vacuum condensate fills all space and provides a background for strong interactions in the cold environment of our cosmic era ("cold" here meaning something less than a trillion degrees Celsius). In order for additional quarks to appear in their colorless clumps of two or three, they must disrupt the condensate in their vicinity. Imagine it has been evaporated to create a little hole, a "bag," in which the excess quarks move and exchange their own mess of gluons and quark/anti-quark pairs. The ubiquitous condensate presses in upon them like water around a bubble. Because the energy needed to create a quark excitation in the condensate includes the energy required to evaporate the interior of the bubble, its mass appears to be much larger than if the condensate were not present. This in fact appears to be the origin of nearly all the mass of things made of quarks (e.g., you and me) in our cold epoch of cosmic evolution. The Higgs mechanism, discussed below, contributes a small starting mass, and the rest is added by interaction with the gluon–quark condensate. In other pictures, the "bag" is replaced by a more complicated distortion of the fluctuating background condensate. In Brookhaven's Relativistic Heavy Ion Collider, for the first time we have an instrument that can heat up and perhaps evaporate the condensate so we might glimpse the "real" quark masses.[20] The even more energetic LHC at CERN may also help unravel this confusing nuclear machinery.

Confinement. Imagine trying to detect two neighboring quark excitations within the QCD vacuum (they would appear to us as a meson). What can we expect to see as the detectors are moved apart? (This is a thought-experiment – we have no detectors that can register the excitations directly.) The likelihood of registration should be greatest when the detectors are close to each other because that configuration requires the least energy in the gluon-binding field. If quarks were particles, we would say that the gluon force pulls them together as closely as possible against the quantum pressure. As the detectors are separated, the probability of registrations will die away rapidly. To get registrations at greater separations we would have to add energy to push the excitations apart against the binding gluon force – think of energizing the electron in the (obsolete) Bohr picture of a hydrogen atom to put it into a larger orbit around the proton. But here we have to add a *lot* of energy because the color force is so strong. At some point, the additional energy needed to see registrations at larger separations exceeds the energy needed to make a brand new pair of quarks. Possibly the energy will go into the gluon field to let the quark registrations appear at even greater separation, but it takes a lot of additional energy for a small gain in distance. It is more likely that the added energy will go into creating the mass for a new quark/anti-quark pair, one close to each of the original ones. So what we would see is four quarks in two separate bundles around the original excitation sites, and much smaller energy in the gluon field. That looks like two mesons.

How far apart can you drive two quarks this way before the binding stuff snaps and creates two new quarks? About the size of protons or neutrons. The quarks are effectively *confined* to this scale. This is the ultimate explanation for the particulate structure of nuclear and chemical matter. Neither the wave functions, nor the fundamental equations of motion, nor plausible initial conditions reveal any hint of atomicity. The atoms we see in today's powerful microscopes occur because stable excitations of quark stuff have an intrinsic tiny size s determined by their mass m_q and the strength F of the gluon force. Pulling quarks apart distance s against force F requires energy Fs. The size s of nuclei can be estimated by setting this equal to the energy of two quark masses. Roughly speaking, $s \approx 2m_q c^2/F$. Once a stable nucleus of this size has formed, ambient electrons float down the electrostatic energy ladder to clothe the confined quark stuff in the oversized electronic garments of chemical atoms.

Asymptotic freedom. The other side of this coin of confinement is that two close quarks interact more weakly than two distant ones because (roughly speaking) there are fewer intervening gluons to enhance the interaction with their own charges. Since the gluons are charged themselves, they in turn excite even more color-charged gluons and also quark/anti-quark pairs. The pieces of this intermediary field are attracting each other as well as binding the two quarks and pushing against the confining condensate. The quark–gluon binding stuff tends to squeeze down into a tube connecting the interacting quarks. Consequently, the color force is strongly directed and concentrated compared with the analogous electric force whose field spreads out widely and so diminishes with distance according to the inverse square law. *The color force does not diminish but remains roughly constant as the quarks move apart!* Calculations estimate the constant tension in the tube of force connecting a quark and an anti-quark at the weight of about 16 metric tons (roughly the tension under which a 1 cm diameter rod of mild steel begins to fail).

At short range – a few tenths the size of a proton – this constant force drops off rapidly, and quarks move almost as if they were free from mutual forces, a phenomenon called *asymptotic freedom*.[21] Thus the strong force behaves with distance just the opposite of the macroscopic electric or gravitational forces. Close quarks move freely, their energy of interaction increasing with distance to a definite limit – a sort of breaking point beyond which new pairs snap into existence to create the illusion of confinement. A similar change of interaction strength on close approach occurs in quantum electrodynamics, but the apparent electric charge grows *stronger* between close charges. The reason is that the mess of photons and electron–positron pairs that are produced between electrically interacting particles have zero net charge and are not confined to a tunnel through a condensate. Thus the strong color force grows weaker, and the electromagnetic force stronger on close approach. Although not required by the Standard Model, it appears as if the two forces might become equal at some very small distance (corresponding to large energy). A future Grand Unified Theory (GUT) "ought" to have a single strength for all the forces and a single multiply dimensioned internal space whose symmetry determines *all* the forces through the Weyl/Yang–Mills mechanism. All such theories to date predict phenomena that have not been discovered, or predict relations among known phenomena that contradict the observations. The concept of a GUT is very attractive, however, and motivates much current research.[22]

Role of pions. What happened to Yukawa's idea that the range of nuclear forces is small because the strong force is mediated by a massive field? Recall that acceptance of that idea damped enthusiasm for the Yang–Mills theory which appears to require massless bosons. Now we know the color-charged gluon force acts as if it had a short range despite having no mass. It is roughly the range over which you can pull quarks apart before a new pair pops into existence. The pion, which everyone thought for decades was the strong force carrier, turns out to be just a quark/anti-quark pair – the lightest colorless hadronic object, and an important but short-lived component of the complicated gluon–quark condensate. The quark gluon field mediating the force among protons and neutrons in the nucleus will manifest itself in colorless pions. Thus pions are indeed associated with the strong nuclear force, but not quite in the way anyone expected. Pions in this picture somewhat resemble *phonons* – quanta of ionic vibrations in a crystal lattice – that can mediate forces among electron excitations in ordinary solids (see below). Here the pions can be regarded as mediating the force not among the elementary quarks, but among the clusters of quarks we call protons and neutrons. So Yukawa and Yang–Mills are both right, but in different domains.

7.6 THE WEAK FORCE (BUT NO QWD)

Electromagnetism is weaker than the strong color force, but the *weak force* is weaker still. It shows up most obviously in the *beta decay* of the neutron (Chapter 6). Half a swarm of free-floating neutrons decays in about 15 minutes – the neutron *half-life*. This is millions of times longer than it takes an excited atom to radiate a photon, which means the neutrons interact much more weakly with their decay products than electrons do. Each neutron seems to break up into a proton, an electron anti-neutrino, and an electron. Neutrons and protons are made of quarks, so beta decay is the result of a force that connects quarks to leptons in the equations of motion. What actually happens is that a d quark in one of the neutrons first slowly radiates a quantum (a boson) of the weak force and turns into a u quark, then the weak boson itself decays almost immediately into the electron and anti-neutrino. Experiments have observed three bosons that carry the weak force.[23] Their conventional symbols are W^+, W^- and Z^0, and each is similar to the photon except they are exceedingly heavy and carry both weak and electric charge, as indicated by the superscripts. These fields link *any* two partners in all three generations of fermions within the Standard Model.

Neutrons make up more than half the mass of our bodies. Are we decaying along with our neutrons? No. When neutrons are trapped in stable nuclei the binding energy keeps them whole. Nestled down in the nuclear well their energy is less than if they were free and in pieces. The mass of an untrapped neutron is greater than the sum of the masses of the free particles into which it is linked by the weak force, thus it can lose energy by breaking up.

How the weak force works. Figure 7.4 shows the weak charge magnitudes ½ and –½ next to the two partners in each generation of fermions, and +1, –1, and 0 next to the weak bosons, which carry weak charge of their own. These numbers measure the weak charge in conventional units that originate in the old isospin concept, which in turn inherited the language of angular momentum where the "allowed" values are measured in multiples of ½. Here the one-halves have nothing to do with angular momentum. It is an accident of mathematics that the internal symmetry that dictates the weak force resembles the rotational symmetry that leads to the conservation of angular momentum. I will say more about the weak symmetry below. Anti-objects have *both* weak and electric charges reversed. A few examples will help visualize how the weak force works.

Consult Figure 7.4, identifying the objects on it as you read this paragraph. The neutral electron neutrino v_e, having weak charge ½, can shake off a W^+ excitation (weak charge 1) and turn into a negatively charged electron (weak charge –½). Similarly, a d quark can shake off a W^- excitation and turn into a u quark, which is half of what happens in beta decay. In the other half, the W^- gives up its energy and its charge to an electron and an electron/anti-neutrino (weak charge –½), just as an energetic photon A can create an electron–positron pair. Using the chart you can see how all the electric and weak charges add up right in these transformations. Emissions of Z-bosons, like photons (you could call photons A-bosons) do not change either electric or weak charge, but both can decay into particle–antiparticle pairs in processes that we will have to consider in more detail below (*electro-weak unification*). The energies and rates of these transformations are strongly affected by the huge masses of the weak bosons: more than 150,000 times the electron mass and heavier than an iron atom ($m_W = 157,400 m_e$, $m_Z = 178,450 m_e$). The origin of these bosons within the theory, described in the next paragraph, gives no clue that they should have any mass at all. An entirely new

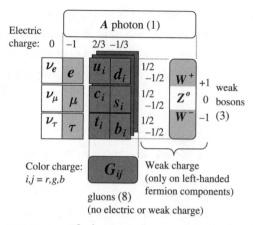

FIGURE 7.4 *The bosons* – photons, gluons, and weak bosons – are shown here in rounded boxes adjoining the fermions to which they are attached in the equations of motion for the wave functions of the Standard Model. All the *left-handed* fermions possess plus or minus one-half unit of weak charge that generates the weak bosons. All fermions but the neutrinos possess electric charge that generates the massless photons *A*, and only the quarks and gluons possess the color charge that generates the massless gluons *G*. The gluons have no electric or weak charge. The massive weak *W* bosons have *both* weak and electric charge. The *Z* boson is very similar to the photon, but massive.

phenomenon is required to explain the masses, described below in the *Higgs mechanism*.

Origin of the weak force. Like all the forces in the Standard Model, the weak force arises from an internal symmetry of Nature: If electricity were turned off, the two members within any family pair of leptons or quarks would closely resemble each other. No doubt much of the mass of an electron compared with its neutrino partner, for example, comes from the energy of the electromagnetic field it generates because of its electric charge. Turn off the charge and it would look very like a neutrino. Same with the up and down quarks. So conjecture that with electricity turned off Nature does not care how we label the pairs. We can change flavor names from one point of space-time to the next. The idea is exactly the same as for isospin, where under the hypothetical no-electricity condition Nature does not distinguish between proton and neutron, so this indifference has the infelicitous name of *weak isospin* invariance. The old isospin is now called *flavor isospin*. Once again we have a legacy of confusing terminology. There is *spin*, which is what you think it is, and two kinds of isospin, only one of which, *weak isospin*, has fundamental significance as an internal symmetry. *Flavor isospin* is a useful but inexact symmetry that describes the close similarity of the u and d quarks and formed the basis of the original Gell-Mann/Zweig quark theory described in Note 10. The only thing all three have in common is their underlying symmetry group $SU(2)$ (Note 16).

In this picture the electron and its neutrino (or any other quark or lepton flavor pair) correspond to "weak isospin down" and "weak isospin up" states measured with respect to some axis in weak isospin space. Nature is symmetric under rotations in this space with the consequence that weak charge is conserved in interactions. Now apply the Weyl/Yang–Mills gauge idea as for the old isospin and discover a new force which must exist that can be "gauge-transformed" to compensate for the arbitrariness in the choice of coordinate axes in weak isospin space about which to measure the rotations. Three new fields are needed, conventionally called W^+, W^-, and W^o, one for each of the generators of rotations in weak isospin space (Note 16). Like the gluons, these new fields can carry the same kind of charge that generates them, namely one unit of weak charge. This corresponds to one unit of weak isospin that can "point" "up" (W^+), "down" (W^-), or "sideways" (W^o) relative to the arbitrary "direction" along which weak isospin is detected. Unlike the gluons and the photon, the weak bosons also carry *electric* charges $+1$, -1, and 0 in units where the electron has charge -1. The superscripts on the W symbols therefore do double duty: they indicate the electric charge and also the weak charge. This is a clue that the electromagnetic and weak forces are somehow intertwined. We have QED and QCD, but we do not have a "QWD" that stands by itself. The theory that treats the weak force consistently is called the *electro-weak* theory.

7.7 ELECTRO-WEAK UNIFICATION

My description of the weak bosons in the preceding paragraph has a curious loose end. It shows up when you ponder how the weak bosons can decay into a pair of leptons. This is going to be a bit technical, but if you trace carefully through the following logic you will glimpse the power of symmetry arguments in this business. It will help to use the notation (electric, weak) to denote the electric charge and weak charge, respectively, of an object. For example, the electron is $e(-1, -\frac{1}{2})$. A little consideration will show that $W^+(+1, +1)$ can only decay into an anti-electron accompanied by a neutrino because the weak and electric charges both must add to plus one. Write the symbols for the decay products this way: $e'(+1, +\frac{1}{2})v_e(0, +\frac{1}{2})$. Similarly a $W^-(-1, -1)$ can only decay to an electron and an anti-neutrino: $e(-1, -\frac{1}{2})v_e'(0, -\frac{1}{2})$. When it comes to W°, however, there appear to be *two* decay modes, where $W^\circ(0,0)$ goes to $v_e(0, +\frac{1}{2})v_e'(0, -\frac{1}{2})$ or to $e(-1, -\frac{1}{2})e'(+1, +\frac{1}{2})$. Neither seems to be favored. Which one does it go to? It had better go to the combination of $v_e v_e'$ and ee' that corresponds to the "sideways" state of a weak-isospin-*one* object, which is what W° is. The wave function for this combination turns out to be $(ee' + v_e v_e')/\sqrt{2}$ using the notation I used earlier for the neutral pion wave function. There is another way of combining a spin up with a spin down to give a combination with *zero* total weak isospin, namely $(ee' - v_e v_e')/\sqrt{2}$. The important point here is that the electron–positron pair ee' *by itself*, without adding or subtracting $v_e v_e'$, *is not in a unique state of weak isospin*, but is rather a combination of states with zero unit and one unit of total weak isospin. Got that?

So what? Well, in QED and in Nature the photon field A generates electron–positron pairs and *vice versa*. That must mean the bosons of A itself are mixtures of states with zero unit and one unit of weak isospin. And remember, "weak isospin" is just a way of describing states with weak charge. This is disturbing because we introduced A as the field that gauges the quantum phase, not rotations in weak isospin space or a mixture of the two. There must be another field, call it B, that gauges the quantum phase, and *the field A we actually observe as the electromagnetic potential must be a combination of B and W°!* A similar argument leads us to conclude that the observed weak neutral boson Z° is also a (different) combination of B and W°. These arguments are actually enough to figure out the right combinations of B and W^0 to make A and Z^0, if only we knew the sizes of the fundamental charges that go with B and the W's.

What we call *electric charge* must also be a combination of the charge that gauges the rotations in weak isospin space (call it g) and the charge that gauges rotations of phase (call it g'). Recall that in QED the phase of the electron–positron wave function is gauged by the product of electric charge q and the photon field A. Using little more than the information in the previous two paragraphs, it is possible to find how q depends upon g and g'. Note 24 gives more detail for the truly dedicated reader.

Summing up, the basic gauge field of QED that I described as the electromagnetic field – conveniently ignoring the weak force – is not the one we observe in actual experiments. That field is a combination of the fields introduced into the theory through the Weyl/Yang–Mills mechanism to account for the invariance of Nature to our choice of reference frames for phase and for weak isospin. Electromagnetism and the weak nuclear force are intimately combined, which is why we now refer to all this as the *electro-weak theory*.[25] Electromagnetism itself is of course the result of a similar merging of the separate concepts of electric and magnetic force in the nineteenth century that began when Oersted noted the deflection of a compass needle by an electric current. How is the vaunted accuracy of QED possible when it omits all reference to the weak force? Because the weak processes only reveal themselves at energies far higher than the usual applications of QED – the threshold energies established by the enormous masses of the W and Z bosons. The mass issue is an urgent one, but before I deal with it, I must catalogue some further odd properties of the weak force.

7.8 PARITY VIOLATION

Of the several odd features of the weak force, the oddest is that *it operates only on fermions with left-handed spin*. This astonishing phenomenon seems to have been suspected by others, but in 1956 T. D. Lee and C. N. Yang acted on their suspicion and sifted through all the experiments that had been done by 1954 looking to see if any of them ruled it out. They came up empty handed, and urged experimenters to look specifically for an asymmetry in beta decay from oriented nuclei. Columbia University's C. S. Wu rapidly orchestrated an experimental team, and sure enough, when the spinning axis of an unstable nucleus is oriented in a magnetic field it spits electrons preferentially in one direction along the axis. The weak force depends on spin direction! *Nature permits left-handed fermions to interact weakly, but not right-handed ones.* Physicists say the symmetry of *parity* is violated.[26]

At a reception in 1995 for members of Stony Brook University's Asian Studies departments, Professor Theresa Kim, an expert in Asian theater arts, spoke to me about a curious phenomenon she had discovered in her ethnological researches. An ancient dance preserved in the Korean shamanistic tradition involved a whirling motion. "Tell me," Professor Kim asked, "Do you know of any phenomenon in nature that involves objects spinning always to the left, not to the right?" Her question left me speechless. Later it occurred to me that if you look up in high northern latitudes into the clear night sky, the stars slowly wheel above as if you yourself were whirling counter-clockwise. Ancient Korean shamans need not have intuited weak isospin to dance in violation of parity. Nature's preference for left-handedness is a mystery deeper than any mystic ever ventured.

To express this phenomenon in the esoteric language I used in the previous section, the left-handed components of each flavor pair of fermions form a weak isospin *doublet*, corresponding to weak isospin $\frac{1}{2}$, that can interact with the W bosons. The right-handed components are *singlets* corresponding to weak isospin zero – i.e., no weak charge at all – and do not participate in the weak interactions.

7.9 CP VIOLATION

Lee and Yang received the Nobel Prize in 1957 only a year after they elucidated the parity question. But 16 years elapsed following the discovery of yet another astonishing property of weak interactions before its importance too was acknowledged with this prestigious award. The issues are somewhat technical (again), but have profound implications (again).

Weak force operates on left-handed fermions and right-handed anti-fermions. If you change the sign of electric charge *and* the direction of spin, right-handed anti-objects turn into left-handed objects. So you might expect the weak force to be invariant under the combined effect of *charge conjugation* (C) and *parity reversal* (P), or CP. That is, the rate at which the u quark, for example, radiates a weak boson and turns into a d quark should be the same as the rate at which the spin-reversed anti-quark u' turns into d'. This can be tested experimentally. In 1964 Princeton University's Val Fitch and James Cronin, working at Brookhaven's Alternating Gradient Synchrotron accelerator, found that CP is nevertheless slightly violated in Nature. The fermion equations of motion (Dirac equations plus interactions as introduced in this chapter) can be compatible with CP violation only when there are at least three generations of fermions (a mathematical fact not easy to visualize), so the effect is *qualitatively* consistent with the Standard Model.[27,28]

This slight violation of CP symmetry implies an asymmetry between matter and anti-matter, which is a good thing if you believe the universe began with a Big Bang that energized all components of matter equally. As the universe cooled, the heavier components presumably decayed into lighter ones through the weak interaction, and among these the condensing objects annihilated each other to produce radiation. The reaction would have proceeded very rapidly in that early dense phase of the universe. If the symmetry between matter and anti-matter were perfect, the annihilations would have consumed all the matter of both kinds, leaving behind only the massless photons and gluons. But we exist, and we are made of matter (mostly u, d, and e, plus bosons), not anti-matter. Therefore, matter must have been created somewhat more abundantly than anti-matter so enough normal matter would remain to make the universe we see today. Unfortunately, the ratio of leftover matter to the leftover radiation, which we can observe, is larger than permitted by the Standard Model. In May 2010 the D0 detector collaboration at Fermilab's Tevatron collider announced measurements of a CP-violating effect observable in the decay of B mesons (b quark bound to s or d quark) into muons that is significantly greater than the Standard Model prediction. "A New Clue to Explain Existence," said the *New York Times*, but as this book goes to press further measurements are needed to confirm the effect.

7.10 THE PROBLEM OF MASS

Our journey into quantum theory began with Maxwell trying to describe in consistent mathematical language a set of straightforward experiments on electricity and magnetism. I have suppressed nearly all the mathematical side of the story, but you can see from the preceding account that technical issues were important in arriving at the Standard Model. Apart from its success in identifying "internal" symmetries that fit the observed multiplicities in the particle spectrum, the Standard Model is the only theory that anyone can find that also is (mostly) mathematically self-consistent. There remain some technically awkward features, but we can still calculate with it (not always easily) and get results that agree (nearly always) with experiments.

Gauge theories of mutual forces among fermions were initially viewed as pretty ideas that could not be correct because the gauge bosons had to be massless, and "everybody knew" that the very short-range weak force had to be mediated by massive bosons in analogy with Yukawa's theory of strong forces. Fortunately, some scientists could not resist exploring the properties of gauge theories simply because they were beautiful. They sought, among other things, clever ways to give mass to the gauge bosons without destroying the gauge-invariance or the renormalizability of the resulting equations. And so the Higgs mechanism was born.

7.11 A DIGRESSION ON SUPERCONDUCTIVITY

The Higgs mechanism is an idea that entered particle theory "the wrong way." Its inspiration came not from meditation upon elementary quantum objects interacting two or three at a time, but from the lore of solid state physics. Within solids, excitations move in a sea of electron stuff that floods the long rows of massive compact ions that make the underlying crystal lattice. The lattice and the electron sea form a background in which waves can move and interact in startling ways. In our theoretical picture of superconductivity, subtle interactions with the lattice link Schrödinger phases among some of the electrons into a rigid quantum structure that can carry current undeterred by random imperfections that otherwise would impede its flow. (Superconductivity may occur in other ways. This ion lattice-mediated interaction is the way we understand best.)[29]

The details are interesting. Normally avoiding each other by mutual electrostatic repulsion, electrons in a solid are never far from oppositely charged ions, which obscure the repulsive force and bend a little toward nearby electrons. The bending changes the environment for other electrons, and thus creates a means for one electron to influence another through local distortions in the ion latticework. This influence amounts to a kind of force among electrons, an effective self-interaction of the Schrödinger electron field, different from the repulsive electrostatic force. In particular, it can be an attractive force, coupling electrons pairwise into units that can be regarded as bosons subject to Bose–Einstein condensation, which changes their collective behavior in a way that stiffens the electron stuff into a super-object. This is an example of the spontaneous symmetry breaking discussed at the end of Chapter 6. The phase of the collective electron wave function locks into a specific value just as the poles of a permanent magnet appear spontaneously in a definite direction as its temperature is reduced below a critical value. In a superconductor, the gauge symmetry of electromagnetism is broken.

Superconductivity is a rare intrusion of the quantum microscopic domain into macroscopic reality. Nature typically reveals herself as mouse-trapped "registrations" in which the strange effects of quantum phase are diluted to inconsequence. Super-currents, however, possess *macroscopic quantum phase*. It is wrong to think of a super-current as a swarm of individual particle-like electrons. The electrons lose their individuality in a uniquely quantum object not easily deflected. They move collectively, as one huge quantum thing, forcing us to recall that "electron" is not a particle, but a phenomenon described by a field – a phenomenon, moreover, whose energy is not limited to the trifling single-particle scales of atoms, but may lumber through a thousand massive magnets, producing in each a small bomb's-worth of energy that must be constrained by sturdy yokes of stainless steel. The super-current in one of Brookhaven's RHIC rings is a quantum beast, alien to our macroscopic world, a Godzilla to be treated with caution and respect.[30]

This huge current is nevertheless a brother to the electron looping an atom in its state of lowest energy. A bound atomic electron is likewise relatively undisturbed by minor perturbations. A passing light wave whose energy is too small will fail to drive it to the next higher "allowed" state. This is a small version of the super-current's stability. The super-current is an electronic state of lowest energy separated from more energetic states by an energy gap. It can be altered only by a disturbance that delivers enough energy to bridge the gap. Normal (non-superconducting) electrons in a metal are separated energetically from their excited states by arbitrarily small energies. They are easily moved from rest. Their response to energizing stimulation has no threshold.

There is something about this behavior of normal and super-conducting electron states that is relevant to the creation of new fundamental objects in particle physics, as first noticed by Y. Nambu in 1960. Accelerated particles smash into targets creating new massive excitations only when the energy of collision exceeds that of the new mass. From the spectroscopic point of view, mass is just the minimum energy needed to excite the system of fundamental objects into a new state. Mass is nothing more than an energy gap in the spectrum of particle excitations.

7.12 THE HIGGS MECHANISM

Perhaps masses – gaps in the particle energy spectrum – appear because some as yet unseen quantum field pervading all space – call it the *Higgs field* – interacts with itself as the superconducting electron field does, lowering its energy into a new macroscopic quantum state and opening up a gap between this new condensed "ground" state and other higher states in the process. To have any effect such a condensed field must interact with other matter, so imagine it is yet another weakly charged flavor pair like the (left-handed) electron/neutrino or up/down quark pair, and thus interacts with the rest of matter through the W or Z bosons of the weak-isospin force. Then when a Standard Model fermion tries to interact weakly with neighboring fermions, it must thrust a weak boson into a world already occupied by the condensed Higgs field. That is to say, not only must it create the boson, but it must also make some disturbance, some dent, in the ever-present Higgs field to which the boson is attached. The energy required for this establishes a threshold for the creation of weak bosons. They appear to have mass.

This is a way to give mass to the weak gauge bosons without putting it explicitly into the equations of motion. In this approach the weak boson fields initially obey *equations of motion* that do not contain mass, but in the presence of the condensed Higgs field the *solutions of the equations* for the motion of the W and Z bosons appear to have mass anyway, proportional to the value of the condensed Higgs field. So there is hope that the whole set of rules (equations) that describe the various interactions will preserve the symmetries needed to make the theory consistent and renormalizable. When you express the idea mathematically you find that the masses of the fermions upset the needed symmetries. So grit your teeth and write the equations to make *all* the fermions interact *directly* with the new field – not just through the weak bosons. Then each time *any* fermion is created, it has to make a dent in the Higgs condensate and it will look like it has a mass. The bigger the interaction strength, the bigger the dent, and the greater the apparent mass – just put in the interaction strengths "by hand" and adjust to fit the observed set of fermion masses. In this way, the condensed Higgs can give mass to *all* the particles observed to possess it. And then step back in astonishment when you find that this desperate recipe actually works!

This is the *Higgs mechanism*. It gives mass to the weak gauge bosons W^+, W^-, and Z^0 but, miraculously, you can arrange it to leave the photon A massless.[31] And it gives mass to all the fermions through the *ad hoc* direct coupling to them. It seems disturbingly like a *deus ex machina*, and indeed it is. But it is the simplest way of getting mass into the picture without spoiling gauge-invariance and renormalizability. And it works. All its predictions have been confirmed, save one. The Higgs field itself has not been observed directly, presumably because its manifestation as a vibration about the condensed ground state requires more energy to excite than our largest existing accelerators can muster.

Peter Higgs, after whom the basic idea is named, made his proposals in 1964, long before the Standard Model rose up out of the mists of theory and particle data. The idea of working the Higgs mechanism into a coherent theory of weak interactions emerged in the late 1960s. Proof of renormalizability finally gave it full status as a viable theory in the early 1970s, and the massive weak bosons were duly discovered at CERN in the mid-1980s. The history is strewn with Nobel Prizes at each step – an indication of the difficulty of the enterprise and its importance to the fundamental understanding of matter.

The Higgs mechanism is a loose end of the Standard Model that drags in a whole menagerie of new issues. What is the origin of the self-interaction that causes the Higgs field to precipitate into a macroscopic "superconducting" state? In a superconducting metal the ion motions are coupled to the electron field. Are there perhaps new as yet unseen objects that link to the Higgs field to make it seem to interact with itself? And how about all these couplings to the fermions? They are all different, very different. Are their magnitudes related somehow to each other or to new forces? Are there additional multi-component objects aligned in undiscovered internal spaces? Why haven't they been observed? How about the energy associated with the spontaneously appearing ubiquitous vacuum Higgs state? Would it not have huge gravitational effects? Why do we not see them? Or do we just not recognize their influence in some aspect of our cosmic environment?

You can see why proud nations might want to invest billions in a machine to probe the Higgs mechanism. It lies at the root of the Standard Model, and holds the key to further understanding of microscopic Nature. It is a surprising, awkward, but absolutely essential part of the best theory we have of Nature. Observing a Higgs excitation would answer more and deeper questions than any previous particle discovery. Which explains the bitter disappointment of particle physicists at the 1993 termination by Congress of the partially completed Superconducting Super Collider near Dallas, Texas.[32]

7.13 THE HIGGS BOSON(S)

Call the Higgs pair ϕ^+, ϕ^o, where the superscript on these Greek *phi*'s indicates electric charge (see Figure 7.5). They are a weak isospin doublet with weak charges $\pm\frac{1}{2}$ like the left-handed fermion partners in each generation of the Standard Model. The weak isospin "up" partner ϕ^+ can shake off a W^+ boson, for example, and turn into the "down" partner ϕ^o. Now the internal weak symmetry makes the direction of "up" and "down" arbitrary. When the self-attractive Higgs condenses, however, it freezes out into a definite weak isospin direction, establishing a preferred frame and breaking the rotational symmetry of the internal environment for all other objects. Use the freedom to choose frames of reference to make "down" along this preferred direction (so the condensed field is electrically neutral). Then the lower component ϕ^o will be the one that exhibits its condensed value which I will call H_o. The apparent masses of the W's and Z are determined by H_o, and the equations of the theory can be worked backward to find its value.

This H_o is the value the Higgs field locks into by virtue of its self-interaction, much as a permanent magnet locks into a certain steady magnetic field strength as it cools. If the system is disturbed, ϕ^o can oscillate about this value. Call the oscillating part H, so $\phi^o = H_o + H$. It is the quantized excitations of H that scientists are trying to observe in collider experiments.

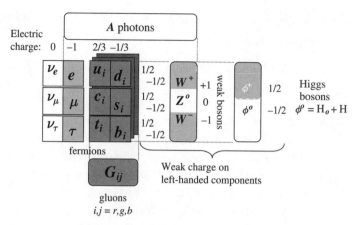

FIGURE 7.5 *The Higgs bosons.* Two Higgs partners are shown here, corresponding to a complete weak isospin pair, but the "up" partner is eliminated by the choice of coordinate axes in weak isospin space. It is conventional to choose "down" to coincide with the arbitrary direction of the Higgs condensate in weak isospin space. The experimentally observable (but not yet observed) object is H, the deviation of the "down" component from its condensed value H_o.

Quantum oscillators with frequency f can only have energies in multiples of hf, so there is a gap of this magnitude in the energy spectrum that can be interpreted as the "rest energy" $m_H c^2$ of a "Higgs particle." Unfortunately, the equations that describe the Higgs theory give us few clues as to the frequency f, and hence the mass m_H. As of this writing (2010), various observations suggest that m_H is just at the upper limit of the Fermilab Tevatron collider, but well within the planned energy of CERN's Large Hadron Collider (LHC) now beginning operation.[33]

From this brief account you can see that the building blocks of the Standard Model are not simply bricks that stack up to form larger structures. They are components of a vast sculptural landscape whose visible manifestations in our era are controlled behind the scene by the pervasive Higgs condensate and the quark–gluon sea. Matter appears as excitations of various omnipresent components within this background, with which it interacts to create the phenomena of massive stable clusters of energy, the stuff of which our worlds are made. There are more components of Nature than the objects that appear in the diagram above, but we have not yet been able to incorporate them in this remarkable framework.

NOTES

1. *Scanning microscopes*. Scanning probe microscopy works by moving a pointed probe just above the sample surface and recording its response to the part of the surface immediately beneath it. Different kinds of probes respond to different properties, and often simply detect an electric current that can flow across the gap between the probe tip and the surface. What is actually measured in this case is the probability of detecting an electron in the region covered by the tip. So the image is essentially a picture of the squared amplitude of the electron wave function. The wave function peaks in the vicinity of atoms where the electrons cluster. Gerd Binnig and Heinrich Rohrer of IBM's Zurich Research Laboratory received the 1986 Nobel Prize in physics for developing this technique.

2. *Scale of nuclear energies*. "If all the atoms in a kilogram of U-235 undergo fission, the energy released is equivalent to the energy released in the explosion of about 20,000 short tons of TNT" (Smyth, 1945). Chemical energy comes from the electrical binding among the charged electrons and nuclei of atoms. Nuclear fission occurs when a stray neutron tips the delicate balance between electrostatic repulsion and the strong but exceedingly short-range nuclear force that binds protons in a heavy nucleus like uranium. The destabilized nucleus deforms catastrophically, torn into two nearly equal pieces by the huge electric repulsion between the parts of its distorted shape.

3. *"Symmetry dictates interaction"*. "We might say that Einstein initiated the principle that *symmetry dictates interactions*. This principle has played an essential role ... in giving rise to various field theories: Coordinate-transformation invariance gives rise to general relativity; Abelian gauge symmetry [under rotations of phase] gives rise to electromagnetism; Non-Abelian gauge symmetry [*e.g.* under rotations in isospin space] gives rise to Non-Abelian gauge fields; Supersymmetry gives rise to a theory with symmetry between fermions and bosons; Supergravity symmetry gives rise to supergravity [not treated in this book]" (Yang, 1983). The word "Abelian," after the mathematician who studied these things, describes a symmetry whose operations may be applied in any order to the same effect.

4. *Other fields and forces*. The measured orbital velocities of stars in galaxies are too rapid to explain by the forces of gravity inferred from the distribution

of visible matter within the galaxies. This suggests the existence of *dark matter* of an unknown origin. Moreover, the observed mutual recession rate of galaxies, attributed to expansion of the entire universe, is greater for nearby (younger) galaxies, so the recession appears to be accelerating as the universe ages. This is attributed to a new phenomenon called *dark energy*. These inferences are supported by other kinds of observations, particularly of the cosmic microwave background "noise" that carries information about the state of the early universe. Many recent books discuss these phenomena at a nontechnical level, e.g., Ferris (1998), Kirshner (2004), Livio (2000), Weinberg (1993).

5. *Three families of leptons.* Joseph John Thompson discovered the *electron* in 1897 at Cambridge University. Its antiparticle, the *positron*, was found among "soft" (less penetrating) cosmic ray debris in 1932 by Carl Anderson at Caltech. The *electron neutrino* was observed finally in 1955 by Frederick Reines and Clyde Cowan from the University of California at Irvine, working at the Savannah River nuclear reactor (reactors produce floods of neutrons and neutrinos). Anderson also discovered the *muon* in the highly penetrating "hard" component of cosmic rays in 1936. The *mu neutrino* was detected in 1962 by Leon Lederman, Melvin Schwartz, and Jack Steinberger at Brookhaven National Laboratory. Martin Perl's group at SLAC (Stanford) found the *taon* in 1975, and its *tau neutrino* was observed at Fermilab in 2000 by the DONUT collaboration, an international group involving scientists from the USA, Japan, Korea, and Greece led by spokesmen Byron Lundberg and Vittorio Paolone.

6. *Limits on neutrino mass.* Neutrinos in the cosmic background are so numerous (about $339/cm^3$) that even with small masses they can influence the distribution of galaxies during their formation. "Neutrinos have a large thermal velocity as a result of their low mass and subsequently erase their own perturbations on scales smaller than the free streaming length ... This subsequently contributes to a suppression of the statistical clustering of galaxies over small scales and can be observed in a galaxy survey" (Thomas et al., 2010). This leads to an upper limit on the masses, a lower limit can be inferred from neutrino oscillations (see Note 8).

7. *Lepton decays.* When an e, μ, or τ lepton decays, it always leaves behind a neutrino from the same generation. When one of these appears as the result of the decay of another particle, it always appears with its anti-neutrino. Thus the muon decays to an electron plus an electron anti-neutrino plus a muon neutrino. Define a *lepton number* for each generation that is "one" for

the leptons, " minus one" for the anti-leptons. Then the sum of the lepton numbers appears to be conserved separately for each generation in all processes. At least it is in all the decays we have observed so far. If neutrinos are their own antiparticles, they could cause reactions that would violate lepton number conservation.

8. *Neutrino oscillations.* Apparently first suggested in a seminar in 1957 at Dubna by the Italian nuclear physicist Bruno Pontecorvo, a student of Fermi's who had emigrated to the Soviet Union in 1950. Brookhaven Laboratory's Raymond Davis was the first to measure the ambient neutrino flux accurately enough to detect a shortfall from theoretical predictions, notably those of Princeton's John Bahcall. Japan's Masatoshi Koshiba also observed neutrinos with a method that permitted their source to be identified as the Sun. Measurements of solar neutrino fluxes provided the earliest information about the masses. The history is summarized in the 2002 Nobel addresses of Davis and Koshiba.

9. *Three families of quarks.* Quarks do not occur in isolation, so they can be "discovered" only through the hadrons made of them. The first theory for fractionally charged hadron constituents was proposed by Murray Gell-Mann and George Zweig in 1964. At that time hadrons made from the three lightest quarks (u,d,s) had been identified in cosmic radiation and accelerators, and it was to account for these observations that the concept of quarks was introduced. The other three quarks (c,b,t) are considerably heavier, and very large accelerators are needed to generate their hadrons artificially. The first experimental evidence for internal constituents within the proton and neutron came from electron scattering experiments at SLAC in 1968 analogous to Rutherford's method for determining nuclear size (see Note 7 in Chapter 3). The first baryon with a heavy quark, the J/Ψ meson (cc'), was discovered simultaneously at SLAC (by a group led by Burton Richter) and Brookhaven (by Samuel Ting's group) in 1974. The bottom quark was discovered in the *upsilon* meson (bb') by Leon Lederman's group at Fermilab in 1977. The top quark does not live long enough to combine with other quarks, so it can be observed only through the pattern of its decay products, which was finally accomplished in 1995 by the two detector teams at Fermilab led, respectively, by the CDF detector co-spokesmen William Carruthers, Jr. and Giorgio Bellittini, and their counterparts for the D0 detector Paul Grannis and Hugh Montgomery. (The names of detectors are usually acronyms – like CDF for Collision Detector at

Fermilab – but D0 is the arbitrary designation of a beam intersection point on engineering drawings of the Fermilab main accelerator ring.)

10. *Color charge*. The idea that a new physical property – "color" charge – might be necessary to account for the observed hadrons was proposed independently by Oscar Greenberg and Yoichiro Nambu in 1964. Without color some observed hadrons would have more than one quark in the same state, which is disallowed by the Pauli exclusion principle (Chapter 5). With color, the quarks can have all their other parameters the same, but differ in color states, "saving" the exclusion principle. It is a confusing accident that the number of *color states* needed to make this work happens to be three, which is also the number of *flavors* (*u*,*d*,*s*) in the Gell-Mann/ Zweig scheme. Some years after he had introduced the word "quark," Gell-Mann claimed to have been inspired by the line "Three quarks for Muster Mark" in a poem in James Joyce's *Finnegan's Wake*, the association with "three" being the key. We now know there are six flavors, not three, but the three-flavor scheme is still useful at energies too low to excite the more massive (*c*,*b*,*t*) quarks.

11. *Hadrons with more than three quarks*. Reported in 2003 at laboratories in the USA, Japan, and Germany. The statistics for these experiments were low (i.e., few events were sighted), and subsequent searches have turned up nothing. The authoritative source for the status of these searches is the biennial *Review of Particle Physics*, available at the Particle Data Group website maintained by the US Department of Energy's Lawrence Berkeley Laboratory (http://pdg.lbl.gov). See Nakamura *et al.* (2010).

12. *Quark masses*. The article on quark masses in Nakamura *et al.* (2010) includes this precaution: "Although one often speaks loosely of quark masses as one would of the mass of the electron or muon, any quantitative statement about the value of a quark mass must make careful reference to the particular theoretical framework that is used to define it. It is important to keep this *scheme dependence* in mind when using the quark mass values tabulated in the data listings."

13. *The polarization of light*. Swedish naturalist Erasmus Bartholinus noted in 1669 that a beam of light entering a certain kind of crystal split into two beams – the two polarizations – a phenomenon for which Christiaan Huygens published a detailed theory (Huygens, 1690). More than a century later, Augustin Jean Fresnel realized (1821) that this phenomenon implied that the optical field, whatever it was, must oscillate *transversely*, i.e.,

perpendicular to the direction of propagation. The two polarizations correspond to the two independent perpendicular directions. This was an important clue to the correct equations describing the propagation of light, which were known well before Maxwell's demonstration that light, in its macroscopic manifestation, was a transverse electromagnetic wave.

14. *Reductionism in biology.* "The pinnacle of fundamentalist reductionism in biology was reached with the Watson–Crick structure of DNA," wrote microbiologist Carl Woese in an important recent assessment. "Molecular biology's success over the last century has come solely from looking at certain ones of the problems biology poses (the gene and the nature of the cell) and looking at them from a purely reductionist point of view. It has produced an astounding harvest. The other problems, evolution and the nature of biological form, molecular biology chose to ignore, either failing outright to recognize them or dismissing them as inconsequential, as historical accidents, fundamentally inexplicable and irrelevant to our understanding of biology. Now this should be cause for pause."

"Twenty-first century biology will concern itself with the great 'nonreductionist' nineteenth century biological problems that molecular biology left untouched. All of these problems are different aspects of one of the great problems in all of science, namely, the nature of (complex) organization" (Woese, 2004).

15. *Feynman diagrams.* Feynman himself gave lucid and immensely popular lectures on QED (Feynman, 1985). His approach is one formulation of a theory developed by physicists after World War II, notably by Feynman himself, Julian Schwinger, Sin-itiro Tomonaga, and Freeman Dyson. Sylvan Schweber's technically detailed history is an authoritative reference (Schweber, 1994). To each line and each vertex in a Feynman diagram corresponds a factor in a mathematical expression. All diagrams in QED can be constructed from electron–positron lines (solid lines below), photon lines (wavy lines), and the fundamental vertex, which can appear in any of its possible topological forms (Figure 7.6)

For example, the diagrams in Figure 7.7 with just two "external" electron lines are a few of the infinite number of terms in an expression

FIGURE 7.6

FIGURE 7.7

that describes the apparent mass of the electron due to interactions with the electromagnetic field (electron *self-energy*).

16. *Symmetry groups SU(2), SU(3), and U(1).* These mysterious abbreviations appear frequently in physics literature. In case you wanted to know, the notation *SU(2)* means "the group of *S*pecial *U*nitary 2-dimensional matrices with complex elements." *Special* means the *determinant* (a measure of size) of the matrix equals one.* *Unitary* means its matrix product with a second matrix obtained from it by interchanging rows and columns and reversing the phase of each element, equals the *unit matrix* which has ones on the diagonal and zeros elsewhere. These restrictions ensure that multiplying any *N*-component wave function by an *SU(N)* matrix will not introduce any spurious phases, and will produce a function that still acts like a probability – i.e., its amplitude does not exceed unity. Without these restrictions, an $N{\times}N$ matrix would have N^2 elements, each of which has a phase and an amplitude, so it takes $2N^2$ numbers or parameters to specify completely. Each restriction on the matrix reduces by one the number of parameters required to specify it (because you can turn that restriction into an equation and use it to solve for one parameter in terms of the others). The *unitary* requirement imposes N^2 conditions (one for each element of the matrix product set equal to one or zero), and the *special* requirement imposes one more, so the number of independent parameters required to specify an *SU(N)* matrix, which is the same as the number of generators, is $2N^2 - N^2 - 1 = N^2 - 1$, or 3 for *SU(2)* and 8 for *SU(3)*. A matrix of the group *U(1)* is just a single complex number whose amplitude equals one so its phase is a single independent parameter. The three generators of *SU(2)* obey the same algebraic rules as the three generators of rotations in ordinary 3-D space, which is why phenomena with this symmetry are called "spins" even when they have nothing to do with ordinary spinning. To carry out the gauge theory, each symmetry group generator is accompanied by a four-vector field that "gauges" how a "nudge" caused by the generator changes from one place in space-time to the next. Thus the *U(1)* symmetry leads to one vector field (the four-vector

potential A_μ in QED, called B_μ in the full theory), the SU(2) symmetry produces three vector fields (describing three weak bosons $W^+_\mu, W^-_\mu, W^o_\mu$), and the SU(3) symmetry produces eight vector fields (the gluon four-vector potentials G^{ij}_μ).

* The *determinant* of an $N \times N$ matrix is the sum of all products of elements taken N at a time, with no two elements from the same row or column, each term in the sum being multiplied by -1 if the order of the elements is an odd permutation of the numbers of the columns in which the elements appear. The determinant appears naturally in the explicit algebraic expression for the inverse of a matrix under matrix multiplication, and for matrices of real numbers is numerically equal to the volume enclosed by the parallelepiped framed by the vectors that form the rows or columns of the matrix.

17. *Gluons*. High-energy collisions sometimes create sprays of particles emerging from the collision in two oppositely directed *jets*. These are interpreted as the creation of a quark/anti-quark pair. As the quarks explode away from each other with the energy of the collision, their confining gluon bonds repeatedly stretch and break to create new quark/anti-quark pairs. Thus a whole string of additional pairs emerges, traveling along the same oppositely directed tracks as the original pair. Sometimes *three* jets are observed, with all three tracks in the same plane, and this is interpreted as evidence of production of an extra gluon along with the quark/anti-quark pair. Observation of such jets at the PETRA electron–positron collider at DESY in Hamburg in 1979 is regarded as the discovery of gluons.

18. *Eight kinds of gluons*. Why not nine? Taking one of the three color charges (r,b,g) and one anti-charge in all combinations makes nine possibilities altogether: rr', rb', rg', br', bb', bg', gr', gb', gg'. The probability of registering a gluon at some point in space will be given by the amplitude of a gluon wave function, and this will generally be a superposition of all nine possibilities. One particular superposition of the nine, however, is totally colorless no matter how you look at it in color space (namely $rr' + bb' + gg'$), and therefore a gluon in that state would not interact with anything, so the state does not represent a physical force. That leaves only eight truly colored combinations.

19. *The QCD vacuum state.* The quark–gluon "vacuum" has a complicated structure that is difficult to infer from the theory. QED-type perturbation theory breaks down here, and it is necessary to solve the equations of QCD on computers to find the state of lowest energy. Specialized computers and computational techniques have been designed for this purpose. Video images of some aspects of the QCD vacuum are available on the World Wide Web. See for example the *Visual QCD* section of the website of the Special Research Center for the Subatomic Structure of Matter (CSSM) at the University of Adelaide in Australia: http://www.physics.adelaide.edu. au/cssm/lattice/.

20. *Relativistic heavy ion collisions.* Brookhaven's RHIC typically accelerates nuclei up to the weight of gold, which has 197 protons and neutrons, and smashes them together at energies of about 200 GeV per nucleon. That's enough energy to create about $200 \times 197 = 39{,}400$ additional nucleons if it were all converted into mass. Most collisions are at oblique incidence but some are nearly head on so the nucleons and their 1182 quarks come to a shuddering halt and the collision energy is translated into a huge number of additional excitations and heat, which melts the surrounding gluon–quark condensate and permits the quarks to interact freely, liberated from their confinement. The idea is to estimate the behavior of this mess of hot matter using computer solutions of the QCD equations and compare with observations of the debris flying from the expanding fireball. The higher energy of CERN's LHC is not required to study this behavior. Visit the Brookhaven National Laboratory website for more explanations and animations (www.bnl.gov).

21. *Asymptotic freedom.* Because the strong force gets weaker as the quarks approach each other, we can expect the approximate calculation techniques successful for QED to work here too. That is why the 2004 Nobel laureates David Gross, Frank Wilczek, and H. David Politzer were able to demonstrate that the particular QCD force law leads to asymptotic freedom. They discovered the remarkable fact that the *only* force laws that can have this property are those derived using the Weyl/Yang–Mills gauge mechanism.

22. *Grand Unified Theories.* See, for example, the brief nonmathematical accounts in Smolin (2007) and Woit (2006) for critiques of extensions of the Standard Model. The simplest extension is based on the symmetry $SU(5)$, which contains the Standard Model symmetry as a subgroup and allows

the proton to decay into leptons. Unfortunately it predicts a decay lifetime too short to be consistent with experiments. While proton decay has never been observed, all extensions of this sort require a finite proton lifetime, which is sufficient justification to keep looking. The idea is to watch for decays in a huge volume of pure water, the large number of nuclei making up for the small probability of decay.

23. *The weak massive vector bosons.* Observed in 1983 at CERN in a series of experiments carried out under the direction of its director Carlo Rubbia, who received in the following year half the Nobel Prize in physics for the feat. The other half went to Simon van der Meer who developed an enhancement to the CERN accelerator that made the observations possible. It was the discovery of a neutral vector boson (Z^o), in particular, that convinced many physicists that the weak interaction must be a gauge force.

24. *Electro-weak forces in terms of gauge fields.* The electromagnetic potential A and neutral weak boson field Z^o can be expressed in terms of the fields B and W^o used to gauge the phase and weak isospin rotations using the gauge charges g, g' as follows:

$$A = (gB + g'W^o)/\sqrt{(g^2 + g'^2)}$$
$$Z^o = (-g'B + gW^o)/\sqrt{(g^2 + g'^2)}$$

if B and W^o were the x and y-coordinates of a vector in a plane, then A and Z^o would be the coordinates of a vector of the same length that was rotated counter-clockwise by an angle θ_W (the *Weinberg angle*) whose trigonometric tangent function is $\tan\theta_W = g'/g$. If the charge on the electron is called $-q$, then $q = g\sin\theta_W = gg'/\sqrt{(g^2 + g'^2)}$. Experiments give $\theta_W \approx 28.1°$, or $g/g' \approx 0.882$.

25. *Electro-weak theory.* Weinberg's own account of the electro-weak theory (Weinberg, 1992) is embedded in a longer discussion of physical theory. Crease and Mann (1996) have a good account of the history of discovery based on interviews with the participants.

26. *Parity.* In view of the great similarity between the families of hadrons and leptons, we might expect parity to be violated in the strong as well as in the weak interactions, but no: "What shocks me," wrote Pauli, "is not the fact that 'God is just left-handed' but the fact that in spite of this He exhibits Himself as left/right symmetric when He expresses Himself strongly. In short, the real problem now is why the strong interactions are left/right symmetric." Quoted in Pais's essay on Pauli (Pais, 2000).

27. *CP violation experiments.* If *CP* were conserved, certain patterns of fermion decay would not be observed, so the search for *CP* violation looks for otherwise forbidden or asymmetric decays. The theory predicts a decay when the wave function of a state of definite flavor evolves into a particular kind of superposition of wave functions of different flavors. The evolution is governed by the relation between the fundamental flavor states and the observable energy states which can be represented by a matrix of complex numbers. This so-called *CKM matrix* (after N. Cabibbo, M. Kobayashi, and T. Maskawa) only permits *CP*-violating decays when its dimension is three or more, which implies the existence of at least three generations of fermions. This insight, whose implication of yet more quarks than the three known at the time was at first disregarded, earned Kobayashi and Maskawa the Nobel Prize in 2008. The absence of *CP* violation in the strong interactions is called "the strong *CP* problem." New fundamental objects called *axions* have been proposed to solve this problem, but none have been discovered.

28. *We exist.* It was Andrei Sakharov, J. R. Oppenheimer's counterpart in the Soviet nuclear weapons program, who pointed out the connection between *CP* violation and the domination of matter over anti-matter in our universe. Sakharov had returned to pure physics research just after Fitch and Cronin reported their observation of *CP* violation, and within a year or two had stated the conditions required for matter/anti-matter asymmetry. Not all the conditions have been shown to occur in Nature, but our own existence, as Descartes succinctly noted, is not in doubt.

29. *Superconductivity.* First observed by the Dutch physicist Heike Kamerlingh Onnes in 1911, superconductivity is a phase of many metals and alloys in which electric currents flow without resistance and magnetic fields cannot penetrate the material. Physicists described the phenomenon mathematically, but could not explain it until 1957 when the mechanism described in the text was proposed by John Bardeen, Leon Cooper, and Robert Schrieffer and is now called the BCS theory. Onnes received the Nobel Prize for his work in 1917, BCS received it in 1972. The span of 46 years between discovery and explanation is very long in the modern era of physics, and even so we do not yet completely understand the kind of superconductivity that occurs at higher temperatures.

30. *Super-current a Godzilla.* In September 2008 a defective connection between two of the 1232 main superconducting magnets in CERN's LHC

ring caused massive destruction in one of the eight sectors of the ring. Each magnet is 15 m long and weighs 30 metric tons. "At the first inspection in the tunnel, many [26] magnets around the two where the defect originated were found to have been displaced with the interconnection bellows heavily damaged ... There was deformation of connections, electrical faults (all but the first induced by mechanical displacement), perforation of the helium vessel [6 metric tons of helium were lost], local destruction of the beam tube with heavy pollution by debris from the electric arc and from fragments of multi-layer insulation ..., breakage or damage of cold support posts, breaches in the interconnection bellows and damage of the warm support jack sustaining the magnets, and cracks in the tunnel floor. The pollution of the beam tubes was much more extensive than the [primary damage] zone, spanning the full 3 km long arc" (Rossi, 2010). The total electromagnetic energy stored in the superconducting magnets was 600 million joules.

31. *No photon mass.* One curious aspect of the Higgs mechanism is that it *requires* a component of the weak force field to be massless. This is awkward because we already have a massless object, the photon, with somewhat similar properties. In a technical *tour de force*, Sheldon Glashow, Abdus Salam, and Steven Weinberg showed how the treatment of electromagnetism and the weak interactions could be combined to produce a single photon-like object (the field A) that is a superposition of the field (B) that gauges the phase and the fourth weak boson (W^o) (see Note 24). Only the combined theories lead to self-consistent results, and therefore we say that the electromagnetic and weak interactions have been *unified*. This work earned its authors a Nobel Prize in 1979. Many others, however, contributed ideas to this technically most intricate of all the parts of the Standard Model.

32. *The Superconducting Super Collider.* During most of the lifetime of the SSC project I chaired the Board of Directors of Universities Research Association, the not-for-profit consortium of research universities with which the US Department of Energy had contracted to build the new laboratory. A useful and accurate brief history of this huge endeavor may be found in Riordan (2001). A more colorful account is embedded in physicist/ author John G. Cramer's science fiction novel *Einstein's Bridge* (Cramer, 1997). See also Steven Weinberg's *Afterword* to Weinberg (1992)

33. *The LHC.* This accelerator, a baryon collider similar to Brookhaven's RHIC or Fermilab's Tevatron but much larger, began operating in 2010 after the initial incident described in Note 30. Each colliding beam will have an energy of 7×10^{12} eV, or 7 Tera-electron volts (TeV) per proton, which is about 7 times the energy of the Tevatron, and 70 times that of RHIC. The collision energy, which is twice this value, is estimated to be well above that required to see evidence of the Higgs phenomenon and hopefully other excitations beyond the Standard Model. The LHC occupies a 27 km circular tunnel originally built in the 1980s for a lepton collider, the Large Electron Positron (LEP) collider. The large size, which served originally to reduce radiation losses from LEP's light electrons and positrons, makes it possible to accelerate baryons (not only protons) to higher energies using superconducting magnets comparable in strength to those used in smaller machines.

8   The proliferation of matter

The Standard Model of matter and its space-time and quantum foundations emerged during the twentieth century in response to a mounting accumulation of empirical evidence. Crucial experiments led to ideas that were refined by ingenious men and women into a logical pattern. Nature, of course, carries this pattern in her bones. Astronomical observations during the past half-century have convinced most scientists that the Nature we can see today is the result of the grandest experiment of all, the very origin and evolution of the universe. It appears as if Nature blazed forth at the beginning of time in the perfect embodiment of a Platonic form only to clothe herself immediately with obscuring veils. The Standard Model is a remnant of that ideal form, but we have been able to piece together at least part of the grand epic of its evolution.[1]

My aim in this chapter is to summarize the principal features of the stable forms of matter of which our everyday world is made – a task that requires only a few components of the Standard Model. But these components are linked to all the others through a series of incredibly brief events at the birth of the universe when, as Steven Weinberg put it, "phenomena directly exhibited the essential simplicity of nature." Completeness demands that I provide some account of these inferred events, however brief.

8.1 AN ABBREVIATED HISTORY OF CREATION

All the energy contained in matter emerged into our universe (we think) in a cataclysmic "Big Bang" about 13.7 billion years ago, an event that energized every component of the Standard Model, plus, it appears, a great deal more. The initial density and temperature of matter were unimaginably great everywhere. Einstein's theory

suggests how the staggering concentration of energy warped the fabric of the universe. The theory suggests an unfurling complex of space, time, and matter that expanded rapidly and cooled with age. It is wrong to think of the matter/energy as bursting from a localized "egg" into surrounding empty space. Endless space itself expanded, each of its points equivalent to every other and mutually fleeing. Along with space, matter everywhere progressively diluted like a design on a swelling balloon, escaping rapidly from that early violence from which everything burst forth.

Our ignorance of physics at very high energy obscures the first explosive instants, but after about one picosecond (10^{-12} s), as the temperature dropped below a million billion (10^{15}) degrees, the machinery of the Standard Model comes into play. At this temperature the characteristic thermal energy is comparable to the heaviest mass-energies of Standard Model components. The facts of the Standard Model, reviewed in the previous chapter, allow us to reconstruct the following sequence of events.[2]

The "essential simplicity" to which Weinberg's eloquence refers is the fact that at extreme temperatures the internal symmetry of the equations of motion is unbroken. Until the Higgs stuff cooled and condensed into its pervasive "vacuum" state, no Standard Model piece had mass – everything behaved like massless radiation with a distribution of energies described by Planck's law (Chapter 3). In the extreme temperatures of the early stages all components were equally present, consistent with the multiplicity of their quantum states (e.g., number of spin directions). As the temperature declined, the Higgs field condensed (and, as well, possibly other undiscovered "symmetry breaking" fields now unseen) giving mass to all particles and altering the effective strengths of the strong, electromagnetic, and weak forces. With further cooling, the declining energy of the radiation could no longer create matter/anti-matter excitations of the most massive fields, and one by one the heavier unstable components disappeared, decaying into lighter objects. Eventually even the lightest, stable, anti-quarks annihilated mutually with (most of) their

normal" counterparts to leave behind a small residue of "normal" baryonic matter and a huge number of massless photons.

The remaining quarks churned in a quark–gluon plasma vastly outnumbered by the lighter leptons and photons. When the universe was about one second old, the temperature had cooled enough for quarks to begin to bind each other and form the first tiny, short-lived clusters of hadronic matter embedded in the restless texture of the gluon–quark "vacuum." Protons and neutrons occurred initially in equal numbers until the thermal energy dropped below their mass difference. The unstable neutrons began to decay but some were captured in the lightest combinations and formed nuclei of deuterium, helium, and a trace of lithium. All this occurred in the first tenth of a millisecond. Astronomical observations reveal other *dark matter* that exists today in vast quantities and contributes to observable distortions of space-time, but otherwise interacts only weakly with the normal matter of the Standard Model. The presence of dark matter is the best empirical evidence we have that the Standard Model is incomplete (except, of course, for its failure to incorporate gravity at all). I am regretfully omitting many important features of this early history. Excellent popular accounts are available that provide much more detail than I can give here.[3]

As far as we know, the universe as a whole is electrically neutral and composed in our era of normal (not anti-) matter, which means the positively charged protons are matched in numbers by negatively charged electrons. In the cooling, thinning environment electron stuff eventually condensed in the vicinity of each nucleus to form the petals, as it were, of the quantum flowers we call chemical atoms. Before this condensation, which occurred about 375 years after the Big Bang, the free electrons scattered the photons as a metal would, rendering the universe essentially opaque to electromagnetic radiation. Once the electrons were bound to nuclei, the photons were decoupled from matter and released to travel throughout space, whereupon their wavelengths continued to increase (and thus energies diminish) in the expanding universe. They appear in our frozen era as a microwave

"backgroundˮ radiation with a Planck spectrum corresponding exactly to a temperature of 2.725°C degrees above absolute zero.

Thus neutralized, the atoms could begin to clump together under the long-range force of gravity, strongly influenced by the dark matter whose electrical neutrality and lack of "friction" with ordinary matter had permitted an invisible clumping at an earlier stage, creating a framework for galactic formation. The initial precipitation of nuclear matter out of the hot soup of quarks and gluons produced nuclei of hydrogen, helium, and a bit of lithium, the three lightest chemical elements. All larger nuclei up to the element iron were forged much later (half a billion years) in the bellies of the great stars – huge clouds of neighboring atoms crunched together by gravity against their quantum pressure, the mutual repulsion of their electrically charged nuclei, and then, at a later stage of compression, against the thrust of nuclear fire. The fire within each star flares up as gravity forces the lighter nuclei to merge and fuse into new combinations with lower energy, the excess appearing as heat and thermal radiation. Gravity is the ultimate victor as the nuclear reactions inevitably run their course, and – if the star has enough mass – the jumble of new elements caves inward in a colossal implosion that forges the remaining elements heavier than iron and recoils magnificently to create a supernova. Thus are the elements cast like seeds into the endless night, chancing rarely, perhaps uniquely, to gather in an hospitable corner of the universe and sort themselves eventually into such as you and me.

Electrons have so little mass that as they cooled and slowed, their de Broglie–Schrödinger waves lengthened far beyond the size of the nuclei to which they were attracted, thus creating a new scale of size for the next, atomic, phase of evolution when the stars began to form. The electrical effect of the electron–nuclear complex is to draw together groups of different atoms into stable clusters or *molecules*. This is the subject of chemistry. If enough different kinds of molecules are available in a form sufficiently dense, and yet sufficiently fluid, some of them can engage in an ongoing chemical reaction that may form multiple copies of themselves. This is the beginning of biology.[4]

Chemical atoms and molecules (nuclei plus electrons) are big enough to see with modern microscopes. They move approximately according to the laws of Newton–Maxwell–Einstein, which are themselves a consequence of the deeper and more elegant machinery of quantum theory. They appear to behave in some respects like "real" atoms." But they are not quite as real as we would wish. Today we understand that all matter obeys quantum rules that cloud the ancient vision of Democritus.

8.2 NUCLEONS AND NUCLEI

As I explained in Chapter 7, the wave functions of the quark clusters called hadrons depend on the locations of the quarks that bind together to make them. Since quarks are fermions, the three-quark hadrons are also fermions. Interchanging the detectors that would register the presence of the constituent quarks must therefore change the sign of the overall wave function. This explains the pattern of plus and minus signs in the formula I mentioned in Chapter 7 for the proton wave function: $p = (u_r u_g d_b - u_r u_b d_g + u_b u_r d_g - u_b u_g d_r + u_g u_b d_r - u_g u_r d_b)/\sqrt{6}$. If you interchange r and g in this expression, the result has the same terms as the original, but with the sign of each term reversed.

Protons and neutrons – nucleons – have no net color charge and so do not attract each other strongly until they are so close their constituent quark wave functions overlap, which happens at roughly the size of the nucleons themselves. Protons repel each other electro-statically, so they will not ordinarily get this close. But violent colli-sions in a hot gas of nucleons (about one hundred million degrees Celsius) bring some of them near enough for strong forces to grab. That is the aim of machines physicists hope may someday produce practical energy from *nuclear fusion.*[5] Nucleons can also clump under the pressure of gravity inside stars. Once they are close enough to stick, some of the clusters will remain stuck (stable nuclei) but most will just scatter, or fall apart spontaneously, or shatter in the violence of colli-sions with surrounding matter.

A proton and a neutron in isolation will just barely stick together to make a stable *deuteron*, but the strong force is not quite strong enough to keep two touching protons together against their electrostatic repulsion. How about two neutrons? They do not stick either, for an interesting reason. The strong force between two nucleons depends on their relative spin direction, and is greatest when the spins are aligned. Binding is strongest when the neutrons are both in the same lowest energy state. Neutrons, being fermions, have vanishing wave functions in this case, so you will not observe them both in the same state unless they have different (opposite) spins, which gives a sub-stantially weaker attractive force. The proton–neutron combination bonds because the constituents are different, so in the deuteron the spins are aligned and the attractive force is stronger. Even so, for every million protons we find in Nature (as the nuclei of hydrogen atoms) we find only about 16 deuterons.

Other combinations of nucleons can be much more stable than deuterium. Each nucleon in a clump can touch multiple neighbors so the force that binds it is multiplied. Helium nuclei (*alpha particles*), for example, have two neutrons and two protons. You can see that the exclusion principle does not prevent them from piling up in the same ground state because each nucleon can still differ from the others through its spin direction or its charge. In general, stable nuclei with even numbers of protons and even numbers of neutrons are much more common than the other possibilities. Nearly 60% of all stable nuclei have this even–even character, and about 40% are odd–even or even–odd (in roughly equal measures), while only about 1.5% have odd proton *and* odd neutron numbers. Nuclei with either protons or neutrons equal to certain "magic" numbers (2, 8, 20, 28, 50, 82, 126) are particularly stable. Helium is "doubly magic" because its proton and neutron numbers both equal two. Other doubly magic nuclei are oxygen-16 $(8n + 8p)$ and calcium-40 $(20n + 20p)$. The exis-tence of these numbers is evidence for a "shell" structure within the nucleus, similar to that of the chemical atoms which I will describe below.

We do not have a periodic table for nuclei, but it is conventional to display nuclei in a table that arrays number of protons (called the *atomic number* Z) vs. number of neutrons (called A). A small portion of this *table of nuclides* is shown in Figure 8.1. Because the chemical properties of atoms are determined by the electric charge on the nucleus, the conventional chemical names and symbols for the elements (H for hydrogen, He for helium, etc.) are just mnemonics for the number Z of protons. Nuclei with the same Z are called *isotopes* of the element with that Z. When writing equations for nuclear reactions, physicists usually indicate the isotope by writing the total number of nucleons as a superscript, so the two stable isotopes of carbon are C^{12} and C^{13}. Notice in the figure that except for H^1 and He^3 all the stable nuclei have more neutrons than protons (or equal numbers), and the heavier nuclei are increasingly neutron-rich. Altogether 237 stable isotopes are known, and another 158 unstable ones have half-lives longer than one year. Every element up to $Z = 83$ (Bi: bismuth) has at least one stable isotope. Heavier elements are generally unstable, but an "island of stability" exists between about $Z = 88$ (Ra: radium) and $Z = 98$ (Cf: californium) where many long-lived isotopes occur of elements such as thorium ($Z = 90$) and uranium ($Z = 93$). Some nuclei, called *isomers*, can live in an excited state for a long time before they settle down to their ground states, emitting radiation in the process. About a dozen of these live longer than a year. Altogether approximately 3000 different nuclei have been observed, most of them very short-lived.

We think of the heavier nuclei as clusters of protons and neutrons like little balls with a thin layer of glue sticking them together, and this picture is useful for many purposes. But nuclei are really assemblies of quarks and gluons, and the sticky ball model exaggerates the persistence of the proton and neutron structures within the nucleus. The strong force among nuclear quarks and gluons makes calculations of nuclear properties from the Standard Model equations of motion difficult, and nuclear physicists are pushing the limits of numerical calculation to glean insights from the theory.

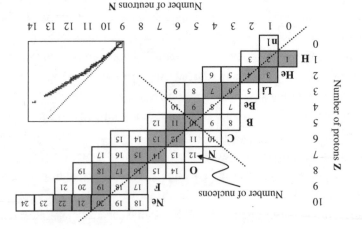

FIGURE 8.1 *A partial chart of the nuclides.* The stable nuclei are shaded, and only a few of the unstable ones are shown as open boxes. Nuclei with the same number of nucleons fall on diagonal lines sloping down to the right. The inset shows all the longer-lived nuclei on a reduced scale. The stable island of a few long-lived isotopes at the heavy end of the mass spectrum includes uranium 238 with a lifetime of about 5 billion years. Each conventional chemical symbol names the whole family of isotopes with the same number Z of protons. (For more detail see the website of the National Nuclear Data Center at Brookhaven National Laboratory: www. nndc.bnl.gov).

8.3 THE PERIODIC TABLE OF ELEMENTS

If you study the chemical elements one by one, from the lightest to the heaviest, you will find a striking periodicity in their chemical properties, first noticed by the great Russian chemist Dmitri Mendeleev in 1869. The properties recur after the first 2 elements, hydrogen and helium, then after the next $2 + 6$ elements for two cycles, then after the next $2 + 6 + 10$ elements for two cycles, then after $2 + 6 + 10 + 14$ elements, and so on. Quantum theory accounts for this regularity in a way that depends critically upon the antisymmetry of the electron wave functions.

De Broglie's electron wave idea established Bohr's "allowed orbits" as approximations to the possible constructive (as opposed to the usually self-destructive) interference patterns an orbiting electron

wave may generate as it hugs the nucleus. Schrödinger's more sophisticated analysis showed that the patterns are much like the vibrations
of a ringing bell. Figure 3.11 shows a pattern that resembles a wave
circling the nucleus, but much more complicated patterns are possible,
some of which are shown in Figure 8.2. These figures show the probability of a registration for a single electron orbiting a nucleus in states
of ever higher energy. They are not pictures of the actual electron stuff:
there is no such picture. The pattern with the lowest energy is a
spherical cloud hugging the nucleus and dying away rapidly with distance. Higher states with definite values of the energy have creases or
nodes where the amplitude vanishes — more nodes for higher energies.
Exactly what the patterns look like depends on how we set the detector
for which the wave function gives registration probabilities, but some
features, like the number of nodes, are characteristic. These are the
features that are correlated with chemical properties. The wave functions of multiple-electron atoms (i.e., all but hydrogen) cannot be
visualized with three-dimensional images, so the following discussion
in terms of one-electron states is an approximation intended only to
illuminate the regularities in the periodic table of elements. It treats
electrons (against all my previous admonitions, and with my apologies)
as independent particles each described by a one-detector wave function unaffected by the presence of the other electrons. While incorrect,
this simplification makes it easier to understand the role of the exclusion principle, or wave function asymmetry, on multi-electron atoms.

As electrons drift toward a multiply charged atomic nucleus, the
first to arrive can snuggle into the lowest spherical energy state, and
the next one can too, if its spin is opposite the first. But the third arrival
will be excluded (by the antisymmetry requirement) because it has no
way to make its wave function different from the first two without
going to a higher energy state. So it settles into one of the states with a
single node in its wave pattern — we say it starts filling a different *shell*
in the layers of electrons that stack up to ever higher energies to avoid
looking like the ones that came before it. The next-lowest-energy
possibility turns out to add a node in the *radial* direction (i.e., the

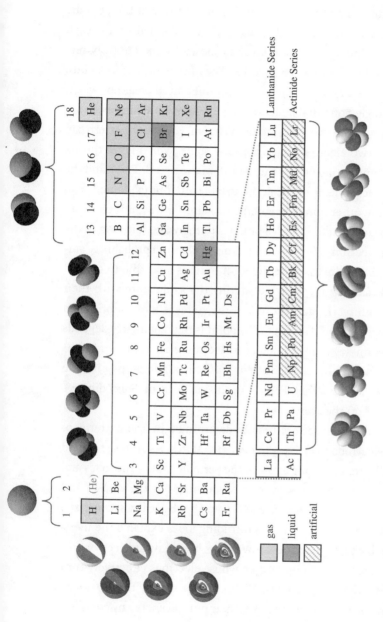

FIGURE 8.2 *The Periodic Table of chemical elements.* Elements in each column, or *group,* have similar chemical properties. The number of electrons in the neutral atom increases from left to right and from top to bottom. All elements after uranium (U) are created artificially in nuclear reactors or accelerators. The lanthanide and actinide series all have the chemical properties of Group 3, which is the first of the ten groups comprising the *transition metals.* Other chemically meaningful names for groups are 1 *alkali metals,* 2 *alkaline earths,* 17 *halogens,* 18 *noble gases.* Groups 13–17 are a mixture of metals, semiconductors, and nonmetals. Probability surfaces are shown for the simplest, or *orbitals,* that have the nodal structure corresponding to the last electron added in the groups below or above them, helium (He) being an exception because both its electrons are in spherically symmetric states. The actual *"electron cloud"* for any group is a superposition of the orbital shown plus all the orbitals for the prior groups. Beyond the surfaces shown, the probability of registering an electron is less than one in 100,000. (Surfaces drawn with *"Orbital Viewer,"* see Figure 3.11.)

node is a spherical surface), so the new shell starts out looking like a layer of an onion. The first two elements in each row of the periodic table (Groups 1 and 2) have electrons in these spherical shells, each new row beginning a new onion layer as shown by the cutaway pictures to the left of each row in Figure 8.2. The fifth electron settles into a pattern that has a *nonradial* node: a circumferential crease or waist. With only a single nonradial node the shape resembles a dumbbell that can point in any of the three directions as shown. Each such pattern can accommodate two electrons with opposite spins, which gives the six columns of elements labeled 13 through 18 in the figure. The 2 and the 6 in the numerical puzzle of the periodic table are thus explained. You can see how the other numbers emerge in a similar way, all related to the possible ways of adding nodes or lobes to the wave function of the next-arriving electron. The small diagrams of electron probability surfaces in Figure 8.2 illustrate the different options available to successive electrons as they try to avoid sharing the same state with their partners. Explaining the actual order of the states is more subtle than this simple picture would suggest. The point is that chemistry depends on the antisymmetry of multi-detector electron wave functions. Bohr somehow conceived the essence of this *Aufbau* (building-up) principle half a decade before either Pauli's exclusion idea or Schrödinger's waves appeared on the scene.

Chemical properties are mostly determined by the patterns of the last few electrons to arrive – the *valence* states. As states are filled within a shell – that is, along a row of the periodic table – the lobes and nodes of their wave patterns overlap, smoothing out the creases and making the completed assemblage in Group 18 spherically symmetric. This smooth neutral ball with no protruding lobes explains the inertness of the noble gases. Each new row begins by adding an electron to a spherically symmetric Group 18 core whose electrons effectively turn their backs to the new arrivals. When atoms touch, their valence electrons arrange themselves to lower their mutual energy by completing the shells on each nucleus, which "explains" many rules of chemical combination.

The simple arguments that "explain" the periodic table do not suffice to calculate the frequencies in the spectrum of light emitted by an electron spiraling toward a nucleus. For that, we need to calculate the ringing patterns of the full wave function, which depends on as many locations as there are electrons near the nucleus. Usually the frequencies can be estimated without working out the full wave function, which is quite complicated. Because such calculations have immense practical importance in chemistry and materials science, scientists have developed clever numerical techniques for doing them efficiently with digital computers.

Complex matter. The table of Standard Model components, the chart of nuclides, and the periodic table of chemical atoms are all comprehensive (except for the dark matter and other undiscovered objects). All the stable combinations of smaller parts, and conceivably all the unstable ones too, can be arrayed in these tabular displays for clusters of elementary components up to the atomic level. But this is as far as we can go. No comprehensive tabulation exists for all the possible stable combinations of chemical atoms. There are just too many possibilities. Once the repulsive electrostatic force among the stable nuclei is nullified by clouds of electrons, the resulting atoms can drift and bind together in countless combinations to produce an awesome variety of chains, rings, spirals, and shapes of every kind. Indefinitely large numbers of atoms can form into molecules, molecules can form clusters which themselves can be units of larger assemblies. Plants and animals are built from snarled chains of amino acids and other molecules that fold and link intricately to perform specific functions within larger organisms. Within this staggering complexity scientists have identified categories in which various kinds of order prevail and lend themselves to systematic study. These are the categories of science itself, arrayed along the reductionist chain that hangs from the Standard Model.

NOTES

1. *The grandest experiment of all.* "To me, the most satisfying thing that has come out of these speculations about the very early universe is the possible

parallel between the history of the universe and its logical structure. . . . The present universe is so cold that the symmetries among the different particles and interactions have been obscured by a kind of freezing; they are not manifest in ordinary phenomena, but have to be expressed mathematically, in our gauge field theories. That which we do now by mathematics was done in the very early universe by heat – physical phenomena directly exhibited the essential simplicity of nature. But no one was there to see it" (Weinberg, 1993).

2. *The Big Bang.* Most likely the appearance of Standard Model components was preceded by a period of *inflation* during which a primordial substance somewhat like the Higgs boson acquired energy that it subsequently released in a Big Bang as it decayed into other species. Many observations point to a Big Bang event, the simplest of which (conceptually speaking) is that the galaxies appear to be fleeing one another at speeds proportional to their relative separations (*Hubble's law*, 1929). This is consistent with a solution of Einstein's equations for gravity in the presence of matter applied to the entire universe. We do not yet have a consistent theory of matter valid for the inflationary period. Popular books by Steven Hawking and Brian Greene (Hawking, 1998, 2001; Greene, 2004) provide glimpses into the issues and some current ideas about how they might eventually be resolved.

3. *The early history of the universe.* Described in lucid detail in Weinberg (1993). The formation and evolution of stars is discussed in a famous popularization by George Gamow, one of the originators of the Big Bang model (Gamow, 1941). Kirshner (2004) is a highly readable account of the history of discovery of the expanding universe, focusing on the recent discovery of its acceleration. See also Greene (2004).

4. *The beginning of biology.* Conditions for the emergence of the complex phenomena of multi-cellular organisms are very difficult to achieve and not likely to occur often in the universe. Ward and Brownlee (2000) is an accessible and systematic account of the conditions under which complex life emerged on Earth.

5. *Practical energy from nuclear fusion.* This has not yet been achieved because it is so difficult to confine matter hot enough for nuclei to fuse. At fusion temperatures, containers made of ceramic or metal would vaporize. Stars contain their fusion fires with gravity. Thermonuclear weapons and concepts for power production called *inertial confinement* reach the critical

temperature only for the fleeting instant of an explosion. *Magnetic confinement* devices attempt continuous confinement with complicated configurations of electric and magnetic fields that act on the charges of the electrons and ions in the plasma, but allow neutral particles to escape. See the website of the international fusion collaboration called ITER: www.iter.org.

Epilogue: Beneath reality

The odd concepts of quantum physics are nearly a century old. They may seem difficult and forbidding at first sight, but the barrier to understanding is not their difficulty but their *differentness*. Unlike classical models, quantum theory does not attempt to simulate Nature. It gives us information about observations of Nature. But it does contain a "theoretical entity" that claims to contain all information, observation notwithstanding, namely the quantum state vector in Hilbert space. Scientists routinely use the language of Hilbert space, and nonscientists do not use it at all, which is understandable but unfortunate. A sort of baby-talk has become the *lingua franca* for much of the popular journalism of quantum physics, an awkward patois that combines some of the early groping language of Bohr, de Broglie, Schrödinger, and Heisenberg with more modern words about symmetries and states. Earnest amateurs still ask me about the mystery of Bohr orbits. There is no mystery because there are no orbits, except as historical curiosities. Same for "wave–particle duality." Same for "quantum jumps." A friend once asked me to address a philosophy seminar about the "paradoxes of quantum mechanics." He said he had approached C. N. Yang first, who replied "What paradoxes?" There are some mathematical and logical rough spots in our current best theory of matter, and perhaps some of them can be framed as paradoxes. But the theory is actually rather straightforward, and it has a perfectly clear interpretation linked to experiments that, in their simplest form, anyone can perform (and students do routinely in college laboratories). No empirical observations are known that are inconsistent with the quantum framework.

It must be admitted, however, that quantum mechanics does fail our expectations in one important respect. It does not specify a unique course of observable events. This is a necessary feature of any theory where the basic descriptive object is a probability function. If a "complete theory" must simulate the course of nature as an orrery simulates the Solar System, then quantum mechanics is not complete. An interpretation of the theory exists, Everett's "many worlds" interpretation, that does make it complete by changing the concept of "reality" to include all the possibilities for which wave functions give probabilities. But this extravagant complexity is too simple. It does not do justice to the seriousness of the conceptual gap that Einstein found so troubling. I think Einstein's deep humanity contributed to his uneasiness. My own sympathies are with Bohr, who seemed more attuned to an unhuman Nature than Einstein ever could have been.

As to the nature of the things around us – our original question – a century's exploration leaves us with the impression that existence, at its core, has a strong geometrical flavor. The metaphors that work best, that lend themselves to mathematical expression and therefore to predictive statements about the world, use the language of spaces and symmetries. We are tempted to say that if only we knew the properties of the geometrical framework in which events of detection may be specified, we would automatically know how the probabilities of observable events within that framework evolve in time. Precisely what it is that causes those events is going to remain elusive. The world's state-vector does not dwell in space and time. Interrogating it leads to phenomena that we call "nonlocal," which we simply cannot picture in our traditional space-time framework.

If we adopt a hard-nosed "show me" view of reality, then the things named in the Standard Model are not real. But everything we can *observe* is real, essentially by definition in common-sense language. *Reality*, in everyday language, is a macroscopic phenomenon. To extend its meaning into the microscopic domain requires a careful, technically narrow, and inevitably subjective definition – inevitable because Nature, the only infallible guide, reveals herself to us macroscopic creatures exclusively in macroscopic events. Empirical data are macroscopic data. The scientific method itself is firmly rooted in the macroscopic world. All we can say of what lies beneath reality is that it is something (I like the word *stuff*) that can exist in different well-defined states, and that can be shaped to trigger macroscopic events with quantitatively computable probabilities.

Those macroscopic events include the formation of spatially confined objects that possess many but not all of the properties of localized lumps of matter – the nuclei of chemical atoms and things made from them – that persist, have a finite size, and that appear to move along rather well-defined trajectories most of the time. Call them particles if you like, but with care. I personally do not like to call the more or less sharp features in the spectrum of energetic radiation "particles," but I confess to doing it so I can be understood by colleagues. But I never like to call the fundamental components of the Standard Model "particles." They are theoretical entities that we use in a mathematical machinery that, among other things, predicts the features of matter's observable energy spectrum.

Physicists perform spectroscopy on matter to identify its different possible states, and find that the states can be correlated with representations of groups of symmetries in abstract spaces of multiple dimensions. The correlation is tricky to perceive because we live in an epoch of cosmic history in which some of the components of matter have frozen out to form one or more background vacuum condensates whose presence masks the underlying symmetry of the whole scheme.

Our successive pictures of that scheme are strongly interconnected: Riemann's speculations on the shape of geometry, followed by Einstein's discovery of the structure of space-time, and gravity as a distortion of its shape, followed by the development of quantum mechanics with its phase like a new dimension in which distortion implies electromagnetic force, followed by the discovery of more Standard Model pieces with new internal spaces in which once again distortions link to nuclear forces. The entire history points to an extraordinary economy in Nature's means.

This growing impression of economy leads us to expect more. We feel that the elaborate Standard Model scheme must be a reflection of something both larger and simpler. The fermion multiplets are almost certainly joined in a representation of a larger, more encompassing symmetry, a *Grand Unified Theory*. A satisfactory theory that describes the phenomenon of gravity in a quantum framework remains elusive, and observations of dark matter in distant galaxies and dark energy in the accelerating expansion of the universe are not yet integrated into the whole picture. Many good ideas about next steps are in the air, but none have yet been brought to Earth by observation. What we do know, moreover, is not completely understood. We hope experiments just beginning at CERN's Large Hadron Collider will point the way. We hope the extraordinary progress in astronomical observations will continue to provide clues. The promise of *string theory*, a tantalizing, immensely difficult not-quite-theory, is likely to remain a promise well into the present century.

We are confident, however, that the quantum framework is correct. And we know that all but one of the pieces of the Standard Model exist in their own inimitable way, because they have been produced and identified in the spectroscopic enterprise of the great accelerator laboratories during the past half-century. These accomplishments are the magnificent legacy of twentieth-century physics to future generations.

Appendix: How quantum mechanics is used

Our large-scale classical view of Nature assumes we can know the current state of things that can be used as an initial condition together with the laws of motion to predict future states (Chapter 2). Newton, Gauss, and others invented procedures for finding the orbits of planets given a few observations of their actual positions in the sky. Since Schrödinger's equation is a law of motion for the wave function, we might attempt similar applications in quantum theory. But how can we know the current wave function of a system? All we can know empirically of Nature, in the quantum view, is whether a detector clicks. This limitation leads to major differences in strategy for using quantum vs. classical mechanics. Keep in mind that the future of one individual atom out of the trillions of trillions in human-sized matter is rarely significant even when the concept of "individual atom" makes sense. By contrast, predicting the future path of a particular near-Earth asteroid may be urgently important. Quantum mechanics is useful despite its ambiguities because we require very different information in the macroscopic and microscopic regimes.

"Particles." The approach to microscopic systems closest to Newton's procedure is to construct or imagine a machine – like a particle accelerator – that repeatedly prepares a system the same way each time. As the machine disgorges successive clones we can observe them with variously tuned detectors. Then from the statistics of the registrations we can attempt to infer a wave function for the system.[1] In Dirac's picture this is equivalent to finding a sufficient number of components of the system's state vector to be confident of its direction in Hilbert space. This approach links the wave function

closely to the device that produces the system. The system's state vector could be said to be a property of the device that prepares it, much as the vector associated with a measurement is defined by the detector. Both devices are macroscopic constructions that define properties of the microscopic states – one in creating, the other in detecting. The operators introduced in Chapter 4 for creating detector coordinates in wave functions – the quantum fields – appear in this new role as operators that create vectors in Hilbert space corresponding to states of Nature that invariably trigger detectors set to certain coordinates. This is how physicists usually think of the quantum fields: as operators that create quantum excitations with well-defined properties. And one of these excitations is precisely what the technical physics term "particle" means today. You can see that this definition is somewhat circular and depends entirely on the ability to construct a detector that always clicks when presented with an object described by such a state. The characteristics of such objects are determined entirely by the symmetry of the geometry in which they are assumed to exist as freely propagating entities, starting with the assumed symmetry of Minkowski's ubiquitous space-time.

The point is that prior to a measurement we simply cannot know the state of a single system unless we prepare it ourselves. Consequently, *calculations in quantum mechanics focus on artificially prepared or hypothetical states.* This is in sharp contrast to the numerous applications of classical mechanics to the motion of "found" objects, from planets and asteroids to baseballs and micro-machinery. The whole point of feedback control in engineering is to use information about the actual present to achieve a desired future. In quantum mechanics we cannot know the present quantum state unless we manufacture it with known machinery. Quantum calculations typically produce information about ideal states that can be incorporated in a larger framework for comparison with experiments. The following examples explain the most important strategies for such calculations.

Energy levels and states. Schrödinger himself introduced a leading paradigm for quantum calculations in his very first paper on wave mechanics. Note 11 in Chapter 4 summarizes his approach. The equation of motion (Schrödinger's equation) relates development in time to wave patterns in space. If you know the pattern now, you can calculate what it will be in the next instant. Schrödinger used his equation "backwards" to examine systems with definite energy. "Definite energy" means the wave pattern in space oscillates with definite frequency, so its time development is known but not its distribution in space. Schrödinger's equation then gives a condition the spatial pattern has to satisfy so as to oscillate with the right frequency. Used this way the equation becomes the *time-independent Schrödinger equation*. There is no guarantee that *any* sensible spatial pattern will oscillate at the right frequency.[2] To appreciate how this works, imagine a tuning fork whose tines vibrate in some pattern linked to a definite tone (frequency) by a classical equation of motion. If you search for a solution of the equation that has a different tone, you will be frustrated. You may find a mathematical solution, but it will have physically impossible features, like tines that move infinitely far from their rest positions. Analyzing the equation of motion will tell you which frequencies will lead to reasonable wave patterns.

In his 1926 paper Schrödinger proposed a wave equation for an electron moving in the presence of a hydrogen nucleus. He assumed their mutual attraction is "given," but otherwise ignored the electromagnetic field. Then he analyzed the resulting equation as described above to find electron wave energies (frequencies) that were compatible with physically possible wave patterns, and found a whole spectrum of them, called by mathematicians the *eigenvalues* of the equation. These turned out to be the same as the "allowed energies" that Bohr had already found in his early model. Corresponding to each eigenvalue is a spatial pattern called the *eigenfunction* for that energy. The wave function corresponding to the least possible energy is called the *ground state*, the others are *excited states*. (See Figure 8.2 for examples.)

This analysis makes no claim that an electron in a real hydrogen atom is ever actually described by any of the eigenfunctions. "Real" electrons are tied to the dynamic electromagnetic field–itself a quantum entity–and are continually buffeted by other atoms, background radiation, and the vacuum condensates of the Standard Model. Their wave functions (or states) are nearly always superpositions of all the different energy eigenstates of the simple equation Schrödinger analyzed. Some of these simple states will contribute more to the "real" state than others. If you could make a detector that responds only to electrons described by the simple states, the probability it will click when tuned to one of the simple eigenvalues is the (squared) amplitude of the contribution to the superposition of the corresponding simple state. Only an electron in an isolated hydrogen atom in its ground state is likely to be described with high accuracy by one of these simple states. Highly accurate measurements show deviations from the simple eigenvalues that can be accounted for by including interactions of the electron with the other Standard Model parts, particularly the lightest ones: photons, electrons, and anti-electrons.[3]

Despite the fact that they rarely describe actual wave functions in Nature, the energy eigenstates of a system, and particularly the energy spectrum, are valuable in other calculations. *Thermodynamics*, for example, is the study of how energy is distributed among the different possible configurations of a system. When a system is in thermal contact with its (presumably much larger) surroundings, the probability it will be found with a certain energy depends only on that energy, the temperature of the surroundings, and on how many different states of the system can have that energy. The exact expression for the probability can be found using statistical methods, assuming the effect of the thermal contact is to randomize the possible states of motion consistent with a constant temperature.[4] Thus information about the energy spectrum and the multiplicity or density of energy states for each energy suffices to determine the thermal properties of physical systems. Detailed knowledge of the wave functions is not required.

The ground state wave function by itself is also useful, especially for multi-component systems where the various pieces tend to arrange themselves in the pattern that minimizes the overall energy. The complex shapes of molecules are usually the shapes of the ground state energy configuration of their component atoms. The precise nature of the ground state is important in other phenomena I described in previous chapters, including superconductivity, Higgs condensation, and the gluon–quark vacuum condensate. Schrödinger's computational strategy may give limited information on real systems, but the spectral information it does provide is immensely valuable for applications.

Transition probabilities and cross-sections. The second fundamental strategy for quantum computations uses the equation of motion for the wave function the way you would expect: start with a known wave and compute how it evolves. The initial "known wave" is often chosen to be an energy eigenstate of a simplified system. That means its corresponding vector in Hilbert space does not point along any of the directions associated with definite energy states of the entire system for which the equation of motion is valid. Consequently the initial state will be a superposition of the energy eigenstates of the whole system. Each component in the superposition oscillates with its own frequency. As time passes, interference will occur among the different components, and some will be suppressed while others grow. The aim is to find the rate of these changes, or the probability of detecting some specific "final" state after the system evolves for a while.

Calculations of this kind are needed to estimate the intensities of spectral lines, and not just their frequencies. They answer the question, for example: If an electron near a nucleus starts in the simple state whose vector in Hilbert space is, say $|i\rangle$, what is the probability it will radiate a photon and end in the final simple state $|f\rangle$ after a time t? To proceed, use the equations of motion to find how the state vector $|i\rangle$ rotates around in Hilbert space to a new direction $|i'\rangle$ at the subsequent time t when $|f\rangle$ is detected, then calculate the projection $\langle f|i'\rangle$. The *probability of transition* is then proportional to $|\langle f|i'\rangle|^2$, which gives the rate photons are produced by this particular transition – proportional to the intensity (rate of detector registrations) of the corresponding spectral line. The point is that we specify *arbitrarily* what the initial and final states are and calculate the rate at which the one acquires a component in the direction of the other as it evolves within the system of interest.[5]

If there is so much arbitrariness in the quantity we compute, how can it describe an actual experiment? One way is to design machinery to produce the initial state and a detector to register the final state. That places both states under our control. Many experiments that use lasers, accelerators, or beams of X-rays, neutrons, atoms, or molecules proceed in this way. An accelerator, for example, creates an object in a particular state with definite energy and momentum, and detectors are arrayed to register the same or other objects that emerge from a *scattering event*. The equations of motion include rules for the interaction between the incoming and scattering objects. The final state may include many new objects that can be detected at various angles relative to the incoming momentum. Transition probabilities in this case are often expressed in terms of a *cross-section* for the event that quantifies our intuition that scattering is more likely from bigger objects.

Transition probabilities among various energy states can also be used to find the intensities in the spectrum emitted by an atom in thermal equilibrium with a heat source. The atoms will be found randomly in various excited states with a probability that can be calculated, as I mentioned above, from independent (nonquantum) statistical considerations. Suppose P_i is the probability of finding an atom in a particular state with energy E_i. The intensity of emission for each possible transition i to f will then be proportional to the rate of change of the transition probability $|\langle f|i'\rangle|^2$ multiplied by P_i. Once again, the quantum result calculated with hypothetical initial and final states is embedded in a larger context that relates the calculation to a realistic situation.

Fundamental properties. In the previous example I assumed we knew the equation of motion including details of the interaction among the incoming objects. In the Standard Model the interactions are given explicitly among the fundamental leptons, quarks, and gauge bosons. But these are *not* the objects we find in laboratories. The actual incoming and detected objects are already complicated combinations of these fundamental pieces. As I explained in Chapter 7, the Standard Model objects are all linked together by the full equations of motion – the Dirac/Weyl/Yang–Mills equations. So we really have two problems: relating the laboratory incoming and outgoing objects to the fundamental objects whose equations of motion we know, and calculating the time evolution (the scattering) of the various fundamental pieces that combine to make the incoming and outgoing objects.

In practice, both problems are handled at once by solving an entirely different problem. We imagine that all the charges, and hence all the forces, are turned off in the distant past and the distant future, and assume the incoming objects are truly fundamental. Then we gradually turn on the forces in the equations of motion, allowing the fundamental objects to interact. The interactions "dress" or convert the fundamental ingoing objects to the actual objects we detect in the laboratory, and they also cause whatever scattering or production of new objects that might occur as the dressed objects interact with each other. In the course of solving the equations of motion (nearly always in some approximation), we carry out the renormalization process that replaces "bare" with "dressed" charges and masses, which *are* the ones we observe in the laboratory (Chapter 4, *Unitarity and renormalization*). For example, we can let the scattering system be an ordinary constant magnetic field and calculate how the electron behaves in it. This tells us the size of the electron's magnetic moment (which agrees closely with experiment). Calculations involving color forces at low energies are much more difficult because they are too strong for the usual approximation methods that work so well for the electro-weak force. At high energies the phenomenon of asymptotic freedom makes weak-force methods applicable.

Despite the remarkable agreement of such calculations with empirical observations, the mathematical status of the relation between the fundamental, i.e., non-interacting, fields and realistic scattered objects is to this day somewhat obscure. The difficulty of inventing a thoroughly rigorous mathematical approach to justify the highly successful calculations of the physicists has preoccupied mathematicians since the early days of quantum theory. The continued elusiveness of their goals has not kept physicists from using the theory anyway, but it has intrigued philosophers of science. Despite our ignorance of the wave functions, and the difficulty in justifying rigorously all mathematical details, quantum theory gives us highly useful information about Nature.[6]

NOTES

1. *We can attempt to infer the wave function.* Since the wave function at any point has both an amplitude and a phase, and only the amplitude can be measured directly (through the statistics of registrations), the problem of inferring the entire wave function is experimentally challenging. It is accomplished for quantum states of the electromagnetic field in a process called *quantum tomography.* The technique requires superposing the unknown state (e.g., an optical beam emerging from some apparatus) with a known state (from a similar apparatus), and detecting a pattern of interference in the registrations of the combined beams. Before each measurement the parameters of the superposition are changed and the amplitude and phase are inferred from the resulting data using mathematical techniques similar to those used in medical imaging devices such as *computerized axial tomography* (CAT) X-ray scanners.

2. *Sensible wave pattern.* What a wave function must look like to be "sensible" has to be decided separately, but usually the requirement is enforced through *boundary conditions.* If the wave function is supposed to describe an electron captured near an atomic nucleus, for example, its magnitude should eventually grow small as you move the detector away from the nucleus. Nearly all choices of energy (in the range of values for which the electron is captured) violate this requirement. The only choices that allow the wave function to die away are the discrete set of eigenvalues of the time-independent equation.

3. *Deviations from the simple eigenvalues.* The first accurate measurement of such a deviation was made by Willis Lamb (1913–2008, Nobel Prize in 1955) and colleagues at Columbia University after World War II (1947). He

measured what is today called the *Lamb shift* of spectral lines arising from the interaction of the electron with the dynamical electromagnetic and electron–positron fields. Calculating these effects is complicated by the fact that the usual perturbation theory produces infinitely large terms (Chapter 6). Applying the renormalization program term by term renders the results finite. Feynman, S. Tomonaga, and J. Schwinger shared a Nobel Prize in 1965 for figuring out how to do this.

4. *The effect of thermal contact is to randomize* ... "Randomize" here means that every possible distribution of the available energy among the energy states of the system and its surroundings is equally likely. This condition implies (through some not too complicated mathematical steps) that the probability of finding the system in a state with energy E increases exponentially with an exponent proportional to the ratio of E to T.

5. *Specify arbitrarily ... initial and final states.* The letters i, i', and f must refer to the state of the electromagnetic field as well as the electron state. The situation described in the text assumes the radiation process begins with no detectable field excitation (i.e., no photons), and an excited eigenstate of the simplified electron plus nucleus system (i.e., without considering the dynamic field). That fixes the initial state. The aim of the calculation is to find the probability that a photon detector will click at time t when tuned to some definite set of photon properties, and at the same time the electron will be found in some definite lower energy state of the simplified system. That fixes the final state.

6. *Difficulty justifying all mathematical details.* Overcoming these difficulties is the aim of a branch of mathematics called *constructive quantum field theory* that seeks, among other things, to explain why the procedures followed by theoretical physicists work as well as they do. I am not aware of any nontechnical treatment of these issues. Even the philosophical literature is highly technical, but the curious reader might try Paul Teller's excellent book (Teller, 1995), especially chapter 6: *Interactions*. The author says the book is written for "interested interpreters of physics" and physics students.

References

Aharonov, Y. and Bohm, D. (1959) Significance of electromagnetic potentials in the quantum theory, *Physical Review*, **115**, 485–91.

Aitchison, I. and Hey, A. (1989) *Gauge Theories in Particle Physics*, 2nd edn. Bristol: Adam Hilger.

Alexander, H. G. (Ed.) (1956) *The Leibniz–Clarke Correspondence*. Manchester: Manchester University Press.

Barrow, J. D. (2000) *The Book of Nothing*. New York: Pantheon Books.

Bell, J. S. (2004) *Speakable and Unspeakable in Quantum Mechanics*, Revised Edition. Cambridge: Cambridge University Press.

Bertlmann, R. and Zeilinger, A. (Eds) (2002) *Quantum [Un]speakables – From Bell to Quantum information*. New York: Springer.

Bohm, D. (1951) *Quantum Theory*. New York: Prentice Hall.

Bohm, D. (1980) *Wholeness and the Implicate Order*. London: Routledge.

Bohr, N. (1913) *On the quantum theory of line spectra*, Kgl. Danske Vid. Selsk. Skr., nat.-math. Afd., 8Raekke IV.1, Part I. Translated and reprinted in Van der Waerden (1967).

Cabrera, B. (1982) First results from a superconductive detector for moving magnetic monopoles, *Physical Review Letters*, **48**, 1378–81.

Capra, F. (1991) *The Tao of Physics*, 3rd edn. Boston: Shambhalla Publications.

Chomsky, N. (1965) *Aspects of the Theory of Syntax*. Cambridge, MA: MIT Press, p. 32.

Clauser, J., Horne, M., Shimony, A. and Holt, R. (1969) Proposed experiment to test local hidden-variable theories, *Physical Review Letters*, **23**, 880–84.

Close et al. (1987). *The Particle Explosion*. New York: Oxford University Press.

Cook, D. J. and Rosemont, H., Jr. (1994) *Gottfried Wilhelm Leibniz: Writings on China*. Chicago: Open Court.

Cooper, L. (1935) *Aristotle, Galileo and the Tower of Pisa*. New York: Ithaca.

Cramer, J. (1997) *Einstein's Bridge*. New York: Eos/HarperCollins.

Crease, R. P. and Mann, C. C. (1996) *The Second Creation: Makers of the Revolution in Twentieth Century Physics*, Revised Edition. New Brunswick: Rutgers University Press.

Davis, P. and Hersh, R. (1981) *The Mathematical Experience*. Boston: Birkhäuser.

De Broglie, L.-V. (1925) *Recherches sur la théorie des quanta*, Ann. d Phys., 10ᵉ série, t. III. Translation by A. F. Kracklauer at http://www.ensmp.fr/aflb/LDB-oeuvres/De_Broglie_Kracklauer.htm.

Dirac, P. (1958) *Principles of Quantum Mechanics*, 4th edn. Oxford: Oxford University Press.

Drake, S. (1957) *Discoveries and Opinions of Galileo*. New York: Anchor Books.

Drake, S. (1981) *Galileo at Work: His Scientific Biography*. Chicago: University of Chicago Press.

Dresden, M. (1987) *H.A. Kramers: Between Tradition and Revolution*. New York: Springer.

Einstein, A. (1905a) On the electrodynamics of moving bodies. *Annalen der Physik*, **17**, translation in Perrett and Jeffery (1923).

Einstein, A. (1905b) Does the inertia of a body depend upon its energy-content? *Annalen der Physik*, **17**, translation in Perrett and Jeffery (1923).

Einstein, A. (1905c) On a heuristic point of view concerning the production and transformation of light, *Annalen der Physik*, **17**, 132–48, translation in *American Journal of Physics*, **33**, 367–74 (1965).

Einstein, A. (1906) The Planck theory of radiation and the theory of specific heat, *Annalen der Physik*, **22**, 180–90 (in German).

Einstein, A. (1925) Quantentheorie des einatomigen idealen Gases, *Sitzungsberichte der Preussischen Akademie der Wissenschaften*, **1**, 3.

Einstein, A. (1916) *Relativity, The Special and the General Theory*, 3rd edn, translated by R. W. Lawson. New York: Crown Publishers (1931).

Einstein, A. and Infeld, L. (1938) *The Evolution of Physics*. New York: Simon & Schuster.

Everett, H. (1957) Relative state formulation of quantum mechanics, *Reviews of Modern Physics*, **29**, 454–62.

Ferris, T. (1998) *The Whole Shebang: A State-of-the-Universe(s) Report*. New York: Simon & Schuster.

Feynman, R. P. (1985) *QED, The Strange Theory of Light and Matter*. Princeton: Princeton University Press.

Feynman, R. and Weinberg, S. (1986) *Elementary Particles and the Laws of Physics*. Cambridge: Cambridge University Press.

Frayn, M. (1998) *Copenhagen*. New York: Anchor Books.

Gamow, G. (1941) *The Birth and Death of the Sun: Stellar Evolution and Subatomic Energy*. New York: Macmillan & Co. (Dover reprint available).

Gauss, C. F. (1827) *General Investigations on Curved Surfaces* [Dover reprint (2005) of an English translation originally published by Princeton University Press (1902)].

Gell-Mann, M. (1994) *The Jaguar and the Quark: Adventures in the Simple and the Complex*. New York: W.H. Freeman and Co.

Gillispie, C. C. (Ed.) (1981) *Dictionary of Scientific Biography*. New York: Charles Scribner's Sons.

Giuliani, D., Joos, E., Kiefer, C., Kupsch, J., Stamatescu, I.-O., and Zeh, H. D. (Eds) (2003) *Decoherence and the Appearance of a Classical World in Quantum Theory*, 2nd edn. New York: Springer.

Greene, B. (1999) *The Elegant Universe*. New York: W.W. Norton & Co.

Greene, B. (2004) *The Fabric of the Cosmos*. New York: Alfred A. Knopf.

Gross, P. and Levitt, N. (1994) *Higher Superstition*. Baltimore: Johns Hopkins, p. 45.

Hawking, S. (1998) *A Brief History of Time*, 2nd edn. New York: Bantam.

Hawking, S. (2001) *The Universe in a Nutshell*. New York: Bantam.

Heath, Sir T. L. (1932) *Greek Astronomy*. London: J.M. Dent & Sons, Ltd.

Heisenberg, W. (1925) On quantum-theoretical reinterpretation of kinematic and mechanical relations, *Zeitschrift fur Physik*, **33**, 879–93 (English translation in V.d. Waerden (1968).

Heisenberg, W. (1930) *The Physical Principles of the Quantum Theory*. Chicago: University of Chicago Press.

Hempel, C. (1966) *Philosophy of Natural Science*. New Jersey: Prentice Hall, p. 83.

Jammer, M. (1966) *The Conceptual Development of Quantum Mechanics*. New York: McGraw-Hill.

Kane, G. (2000) *Supersymmetry: Unveiling the Ultimate Laws of Nature*. Cambridge, MA: Perseus Publishing.

Kant, I. (1783) *Prolegomena to Any Future Metaphysics*, English edition edited by L. W. Beck. New York: The Liberal Arts Press (1950). Remark I to Part 1.

Kepler, J. (1618) *Epitome of Copernican Astronomy Book IV*, reprinted in *Epitome of Copernican Astronomy and Harmonies of the World*, translated by Charles Glenn Wallis. New York: Prometheus Books (1995), pp. 119–20.

Kirshner, R. (2004) *The Extravagant Universe*. Princeton: Princeton University Press.

Klein, M. (1967) *Letters on Wave Mechanics*. London: Vision Press Ltd.

Klein, M. (1980) *Mathematics: The Loss of Certainty*. New York: Oxford University Press, p. 67.

Kline, M. (1985) *Mathematics for the Non-Mathematician*, Dover reprint of *Mathematics for Liberal Arts*. Reading: Addison-Wesley (1967).

Koestler, A. (1963) *The Sleepwalkers*. New York: Grosset and Dunlap, Part V.II.

Kragh, H. (1999) *Quantum Generations: A History of Physics in the Twentieth Century*. Princeton: Princeton University Press.

Laplace, P. S. de (1814) *Essai philosophique sur les probabilities*.

Lindley, D. (1996) *Where Does the Weirdness Go?* New York: Basic Books.

Livio, M. (2000) *The Accelerating Universe*. New York: John Wiley & Sons.

Livio, M. (2005) *The Equation That Couldn't Be Solved*. New York: Simon & Schuster.

London, F. (1927) Quantum mechanical interpretation of Weyl's theory, *Zeitschrift fur Physik*, **42**, 375.

Lorentz, H. A. (1915) *The Theory of Electrons*, 2nd edn. Dover reprint 1953. Note 75*.

Mach, E. (1893) *The Science of Mechanics: A Critical and Historical Account of its Development*, 1st English edition, The Open Court Publishing Co., translated by Thomas J. McCormack, 6th edn. Illinois: Open Court (1960), p. 589.

Marburger, J. (1996) What is a photon?, *The Physics Teacher*, **34**, 482.

Marburger, J. (2008) A historical proof of Heisenberg's uncertainty relation ..., *American Journal of Physics*, **76**, 285.

Marburger, J. and Lam, J. (1979) Nonlinear theory of degenerate four-wave mixing, *Applied Physics Letters*, **34**, 389.

Maxwell, J. C., On physical lines of force, Parts 1 and 2, *Philosophical Magazine*, **XXI** (1861) London.

Maxwell, J. C. (1873) *A Treatise on Electricity and Magnetism*, 3rd edn. Dover reprint 1954.

Mermin, N. D. (1985) Is the Moon there when nobody looks? Reality and the quantum theory, *Physics Today*, **38**, 38–47.

Mermin, N. D. (2005) *It's About Time*. Princeton: Princeton University Press.

Minkowski, H. (1908) *Space and Time*. In Perrett and Jeffery (1923).

Nahin, P. (1998) *An Imaginary Tale: The Story of $\sqrt{-1}$*. Princeton: Princeton University Press.

Nakamura, K. *et al.* (2010) (Particle Data Group), *Journal of Physics G*, **37**, 075021.

Needham, J. (1959) *Science and Civilization in China*. Cambridge: Cambridge University Press.

Neugebauer, O. (1957) *The Exact Sciences in Antiquity*, 2nd edn. Providence: Brown University Press.

Newton, I. (1686) *Principia Mathematica*, translated by A. Motte, edited by F. Cajori. Berkeley: University of California Press (1962).

Newton, I. (1704) *Opticks*. Dover reprint 1952.

Nichol, L. (Ed.) (1998) *On Creativity*. London: Routledge.

O'Raifeartaigh, L. (1997) *The Dawning of Gauge Theory*. Princeton: Princeton University Press.

Omnès, R. (1994) *The Interpretation of Quantum Mechanics*. Princeton: Princeton University Press.

Pais, A. (1982) 'Subtle is the Lord ...': The Science and the Life of Albert Einstein. New York: Oxford.

Pais, A. (1986) Inward Bound: Of Matter and Forces in the Physical World. New York: Oxford University Press.

Pais, A. (1991) Niels Bohr's Times. New York: Oxford University Press.

Pais, A. (2000) The Genius of Science. New York: Oxford University Press.

Pais, A., Jacob, M., Olive, D., and Atiyah, M. (1998) Paul Dirac: the Man and His Work. Cambridge: Cambridge University Press.

Penrose, R. (1989) The Emperor's New Mind. Oxford: Oxford University Press.

Perrett, W. and Jeffery, G. B. (1923) The Principle of Relativity: A Collection of Original Memoirs on the Special and General Theory of Relativity. New York: Dover.

Planck, M. (1900) On an improvement of Wien's equation for the spectrum, Verhandl. Dtsch. phys. Ges., 2, 202. English translation in Haar, D. ter (1967) The Old Quantum Theory, Oxford: Pergamon Press.

Pullman, B. (1998) The Atom in the History of Human Thought. New York: Oxford University Press, chapter 19.

Pyle, A. (1995) Atomism and Its Critics. Bristol: Thoemmes Press.

Quigg, C. (2004) Beyond the Standard Model in Many Directions. arXiv hep-ph/0404228, http://arxiv.org.

Redhead, M. (1995) From Physics to Metaphysics. Cambridge: Cambridge University Press.

Riemann, G. F. B. (1854) On the Hypotheses which Lie at the Foundation of Geometry, English translation by H. S. White in Smith (1959).

Riordan, M. (2001) A Tale of Two Cultures: Building the Superconducting Super Collider, 1988–1993, in Historical Studies in the Physical and Biological Sciences, Vol. 32, p. 125, University of California Press.

Rossi, L. (2010) Superconductivity: its role, its success and its setbacks in the Large Hadron Collider of CERN, Superconductor Science and Technology, 23, 1–17.

Saunders, S. and Brown, H. R. (Eds) (1991) The Philosophy of Vacuum. Oxford: Clarendon Press.

Schilpp, P. A. (Ed.) (1959) Albert Einstein: Philosopher-Scientist. New York: Harper.

Schrödinger, E. (1926) Quantisation as a problem of proper values (Part I), Annalen der Physik, 79, 361. English translation in Collected Papers on Wave Mechanics. Providence: Chelsea Publishing (1982).

Schweber, S. (1994) QED and the Men Who Made It: Dyson, Feynman, Schwinger and Tomonaga. Princeton: Princeton University Press.

Schwinger, J. (Ed.) (1958) Selected Papers on Quantum Electrodynamics. New York: Dover.

Shamos, M. (Ed.) (1959) *Great Experiments in Physics*. New York: Holt, Rinehart and Winston.

Smith, D. E. (1959) *A Source Book in Mathematics*. New York: Dover.

Smolin, L. (2001) *Three Roads to Quantum Gravity*. New York: Basic Books.

Smolin, L. (2006) *The Trouble with Physics*. New York: Basic Books.

Smyth, H. (1945) *Atomic Energy for Military Purposes*. Princeton: Princeton University Press.

Teller, P. (1995) *An Interpretive Introduction to Quantum Field Theory*. Princeton: Princeton University Press.

Thomas, S., Abdalla, F., and Lahav, O. (2010) Upper bound of 0.28 eV neutrino masses from the largest photometric redshift survey, *Physical Review Letters*, **105**, 031301.

Van der Waerden, B. L. (Ed.) (1968) *Sources of Quantum Mechanics*. Dover reprint.

von Neumann, J. (1955) *Mathematical Foundations of Quantum Mechanics*. Princeton: Princeton University Press (English translation of the German edition, Springer, Berlin, 1932).

Ward, P. and Brownlee, D. (2000) *Rare Earth*. New York: Springer.

Watson, J. (1968) *The Double Helix*. London: Atheneum.

Weinberg, A. (1967) *Reflections on Big Science*. Cambridge, MA: MIT Press.

Weinberg, A. (1961) Impact of large scale science on the United States, *Science*, **134**.

Weinberg, S. (1992) *Dreams of a Final Theory*. New York: Pantheon.

Weinberg, S. (1993) *The First Three Minutes: A Modern View of the Origin of the Universe*, Updated Edition. New York: Basic Books.

Weinberg, S. (1995) *The Quantum Theory of Fields: Volume 1, Foundations*. Cambridge: Cambridge University Press.

Weinberg, S. (2003) *The Discovery of Subatomic Particles*, Revised Edition. Cambridge: Cambridge University Press.

Weisskopf, V. (1968) The three spectroscopies, *Scientific American*, **May**, p. 15.

Weyl, H. (1930) *The Theory of Groups and Quantum Mechanics*, 2nd edn. Dover edition of English translation 1950. Original German edition 1928.

Weyl, H. (1952) *Symmetry*. Princeton: Princeton University Press.

Wheeler, J. A. (1998) *Geons, Black Holes, and Quantum Foam: A Life in Physics*, with Kenneth Ford. New York: W. W. Norton & Co.

Wheeler, J. A. and Zurek, W. H. (Eds) (1983) *Quantum Theory and Measurement*. Princeton: Princeton University Press.

Whittaker, E. T. (1910) *A History of the Theories of the Aether and Electricity*, Vol. I. London: Longman.

Wigner, E. (1959) *Group Theory and its Application to the Quantum Mechanics of Atomic Spectra*. New York: Academic Press. Original German edition 1931.

Wilford, J. N. (2000) *The Mapmakers*, Revised edition. New York: Alfred A. Knopf.

Williams, L. P. (1965) *Michael Faraday*. New York: Basic Books.

Woese, C. (2004) A new biology for a new century, *Microbiology and Molecular Biology Reviews*, **June**, 173–86.

Woit, P. (2006) *Not Even Wrong*. New York: Basic Books.

Wu, T. T. and Yang, C. N. (2006) *International Journal of Modern Physics A*, **21**, 32–5.

Yang, C. N. (1983) *Selected Papers 1945–1980 with Commentary*. San Francisco: W.H. Freeman & Co.

Yang, C. N. (2003) Thematic melodies of twentieth century theoretical physics: quantization, symmetry, and phase factor. *International Journal of Modern Physics A*, **18**, 1.

Zajonc, A. (Ed.) (2004) *The New Physics and Cosmology: Dialogues with the Dalai Lama*. Oxford: Oxford University Press.

Zukav, G. (1979) *The Dancing Wu Li Masters*. New York: William Morrow & Co.

Index